T0176663

Energy and Global Climate Change

Energy and Global Climate Change

Bridging the Sustainable Development Divide

Anilla Cherian

WILEY Blackwell

Contents

Foreword

The history of the 21st century may well be determined by our ability to confront the threat of climate change while bringing the benefits of development to all citizens of the world. There is a growing hope that we can accomplish these two linked objectives. Countries are presently working toward a new climate agreement and a new set of sustainable development goals, with aims to conclude both sets of negotiations in 2015. With the affordable, scalable solutions currently available, there is a real opportunity of leapfrogging to cleaner, more resilient economies, providing prosperity and security for this and future generations. The pressing question is whether we have the political courage to bring to a successful conclusion two complex sets of negotiations, each with very high stakes and a decades-long history characterized by swings between optimism and skepticism.

Climate change is not a far-off problem. It is happening now and is having very real consequences on people's lives. As the Intergovernmental Panel on Climate Change (IPCC) reminded us in its *Fifth Assessment Report*, our influence on the climate system is clear and emissions of greenhouse gases (GHGs) are the highest in recorded history. The gradual global warming and increased climate variability have had widespread impacts on both human and natural systems – the atmosphere and ocean are warming, snow and ice levels are diminishing, and sea levels are rising. Many of these changes are unprecedented. The IPCC is unequivocal that, without additional efforts, warming by the end of the 21st century will lead to severe, widespread, and irreversible impacts globally.

There are different ways to achieve the required GHG emissions reductions, but all of them involve changing the world's energy system. In the developed world, energy is almost universally available and accessible. Illumination at the flick of a switch, heat for cooking, and comfort at the turn of a knob – these are taken for granted in developed countries and are hardly given a thought, except on those rare occasions when our sophisticated but mostly unseen energy systems fail. In many developing countries, the picture is very different. It is a harsh reality that in an increasingly hyper-connected world, 2.4 billion human beings must still rely on biomass to meet most of their energy needs and some 1.4 billion people have no access to electricity.

This means millions of people – mainly women – spend hours each day in the arduous business of collecting firewood, animal dung, or crop residues to cook or to heat their homes. Burning these fuels exposes their families to pollutants that cause chronic respiratory and eye diseases, disproportionately affecting women and their young children. This results in millions of deaths each year. Lack of access to electricity and the services it makes possible retards efforts to improve health and education and to provide clean water and sanitation services.

Enabling the world's poor to have access to energy is a development priority. The good news is that providing modern energy services to every person on the planet need not dramatically increase the emission of GHGs and thereby add to the climate problem. We have a very real opportunity to act in a way that is "climate-wise," laying the foundations for a low-carbon-energy development path rather than following a business-as-usual fossil trajectory. Such an approach would bring additional benefits – more jobs, reduced imports of fossil fuels, improved air quality and other local environment benefits, and improved health – that come from the use of low-carbon-energy technologies that are available now and that continue to drop in price. This will require concerted international action. Countries with low levels of access to modern energy services – mainly in Africa and South Asia – will need both financial and technical resources to address the challenge; providing this assistance *must* be a high priority outcome from global negotiations on both the sustainable development goals and climate change.

Recognizing the benefits of low-carbon development and the importance of energy in meeting development goals, in 2013 UN Secretary-General Ban Ki-moon launched an initiative called "Sustainable Energy for All". SE4ALL is an investment in our collective future. Its three goals, all to be reached by 2030, focus on providing universal energy access, increasing the use of renewable energy, and improving energy efficiency, all while addressing the nexus between energy and health, women, food, water and other development issues that are of interest to all countries. Sustainable energy provides new opportunities for economic growth. It enables businesses to expand, generates jobs, and creates new markets. With sustainable energy, children can study after dark. Clinics can store life-saving vaccines. Countries can grow more resilient to future climatic changes, and build the competitive, robust economies of the future.

Despite tremendous progress, barriers still prevent wider adoption of sustainable energy solutions. More effort is needed in areas such as finance, technology development and dissemination, policy and regulatory innovation, and improved business models and governance structures.

With this in mind, in September 2014 the Secretary-General invited world leaders, from government, finance, business, and civil society to New York to galvanize and catalyze climate action. He asked these leaders to bring bold announcements and actions to the Summit that would reduce emissions, strengthen climate resilience, and mobilize political will for a meaningful international legal agreement in 2015. The Climate Summit provided an opportunity for leaders to champion an ambitious vision, anchored in action that supports a strong global agreement in Paris. The Summit made clear that the benefits of taking action to address climate change in a manner that recognizes development priorities such as clean, affordable energy have become ever more compelling, and the need for action more urgent. With recent developments, we have cause for optimism.

Energy and Global Climate Change: Bridging the Sustainable Development Divide argues in a powerful and insightful manner that intergovernmental discussions aimed at solving the climate change and energy access challenges have suffered by being placed in different negotiating silos. This compartmentalization has hindered mutual understanding and made it more difficult to find areas of compromise and agreement on actions that need to be taken, including a better appreciation of the potential "win–win" opportunities. Anilla Cherian's analysis presents the linking of these agendas as the means of breaking these twin impasses and moving forward with a more integrated climate and development approach.

Some may argue that negotiations among governments about these two imperatives are difficult enough without conflating separate streams of discussion and debate. But even advocates of separation will appreciate the author's analysis of the history of intergovernmental negotiations regarding climate change and energy access in the context of sustainable development, and her persuasive calls for a broader, more holistic approach.

Achim Steiner
United Nations Under-Secretary General
Executive Director, United Nations Environment Programme

Acknowledgments

This book would not have seen the light of day without the love and support of my family – my parents, Elizabeth and Abraham, and especially my husband John, and our sons, Rohan and Arman.

I owe a debt of gratitude to Achim Steiner, the Executive Director of the United Nations Environment Programme (UNEP), for his compelling foreword which summarizes the rationale and context of the dual, intersecting challenges of climate change and energy access. Under his leadership, UNEP has been at the forefront of policy and research on the importance of addressing the linkages between the lack of access to sustainable energy services for the poor and climate change. I am also very grateful for the expert inputs of Mark Radka of UNEP. Having worked on energy, climate change, and sustainable development issues for close to two decades, there are many friends, colleagues and experts whose insights, wisdom, and friendship I have benefited from through the years. I would, however, like to single out Prof. Youba Sokona and Prof. Peter Haas for their valuable comments during the review process.

The initial support for my book proposal by Wiley's former Executive Commissioning Editor, Rachael Ballard, was a welcome respite as this book was written under adverse personal health conditions. After a detailed review process, in August 2013 Wiley's editorial team, comprising Rachael, Lucy Sayer, and Fiona Seymour, decided to take a chance on a book written by an independent energy and environment researcher that focuses on one of the most analyzed and polarizing global environmental issues of our times – global climate change – and its linkages to the lack of energy access for the poor within the framework of UN global outcomes. I am grateful for my Managing Editor, Delia Sandford Oxford's constant support, and for inputs from Ian Francis and Audrie Tan, and, of course, I am especially thankful to have a wonderful copy-editor, Adam Campbell. It was important to have an adequately representative cover. The cover image is a painting that I did many years ago that incorporated twigs and tree bark which serve as stark reminders of the daily energy sources used by millions who lack access to modern energy services. Thanks are owed to Yvonne Kok for the cover design, and to Mahenau and Sandra for their suggestions.

In the end, the impetus for the book is the idea that addressing linkages rather than silos is critical for dealing with complicated yet intersecting global development challenges such as climate change and energy access for the poor. The global acknowledgment of energy as a shared driver of sustainable development and global climate change makes the existence of two, separate UN negotiated outcome silos on energy for sustainable development and climate change hard to reconcile, and harder still to ignore. The UN and its member states have resolved to undertake ambitious global action to reduce poverty and mitigate climate change to ensure a sustainable future for all. Responding to the poverty-energy-climate change nexus necessitates integrated global approaches that can effectively address issues such as lack of energy access for the poor, as well as, greenhouse gas emissions and short lived climate pollutants. Concerted global action to develop a new generation of sustainable energy services that simultaneously address global climate change and the needs of the energy poor is imperative. The year 2015 is a landmark year for addressing complex global sustainable development challenges within the UN, but continuing to respond to two intrinsically linked development issues – energy access and climate change – within two separate UN-negotiated silos makes little sense given the UN-led quest for an integrated and universal post-2015 development agenda.

Anilla Cherian
May 2015

1

Confronting the Neglected Nexus Between Climate Change and Energy Access for the Poor

As I see it, we face two urgent energy challenges. The first is that one in five people on the planet lacks access to electricity. Twice as many, almost three billion, use wood, coal, charcoal or animal waste to cook meals and heat homes, exposing themselves and their families to harmful smoke and fumes. This energy poverty is devastating to human development. The second challenge is climate change. Greenhouse gases emitted from burning fossil fuels contribute directly to the warming of the Earth's atmosphere, with all the attendant consequences: a rising incidence of extreme weather and natural disasters that jeopardize lives, livelihoods and our children's future. Sustainable energy for all by 2030 is an enormous challenge. But it is achievable.... We need to raise sustainable energy to the top of the global agenda and focus our attention, ingenuity, resources, and investments to make it a reality.

Ban Ki-moon, 2012, *New York Times*

1.1 Confronting the neglected nexus between climate change and energy access for the poor: Time for "bold action"

The issue of global climate change is not new to the intergovernmental arena. It first emerged in 1988, when the UN General Assembly (UNGA) adopted a resolution sponsored by the Government of Malta, recognizing climate change as a "common concern of mankind" (UNGA, 1988). In the ensuing years, global climate change has been viewed as one of the most complex socio-economic and

Energy and Global Climate Change: Bridging the Sustainable Development Divide,
First Edition. Anilla Cherian.
© 2015 John Wiley & Sons, Ltd. Published 2015 by John Wiley & Sons, Ltd.

political challenges facing the world, and also as one of the most complicated and politically fraught global environmental problems to assess and negotiate (Stern Review, 2006). Climate change has been defined by the first, and to date the only, framework agreement on the subject, the United Nations Framework Convention on Climate Change (UNFCCC), as "a change of climate which is attributed directly or indirectly to human activity that alters the composition of the global atmosphere, and which is in addition to natural climate variability observed over comparable time periods" (Information Unit on Climate Change [IUCC], 1992, p. 5). In this context, the impacts of climate change can be viewed as the existing and potential impacts associated with increases in anthropogenic greenhouse gas (GHG) emissions that enhance the Earth's natural greenhouse effect and result, on average, in an additional warming of the Earth's surface and atmosphere. According to the UN, some of the main characteristics of climate change are increases in average global temperature (global warming), changes in terrestrial cloud cover and precipitation, melting of ice caps and glaciers and reduced snow cover, and increases in ocean temperatures and acidity due to absorption of atmospheric heat and carbon dioxide (CO_2).

Climate change is anticipated to have wide-ranging effects on a variety of socio-economic sectors and areas, including water resources, agriculture, food security, human health, terrestrial ecosystems and biodiversity, oceans and coastal zones. The UN predicts that billions of people, particularly those in developing countries, are highly vulnerable to adverse climatic impacts and will face shortages of water and food and greater risks to health and life as a result of climate change, because they have fewer resources with which to adapt (UNFCCC, 2008). Climate change has been estimated to compound and exacerbate existing poverty because its adverse impacts are more keenly felt amongst the poor in developing countries, given their dependence on fragile and vulnerable ecosystems and natural resources, as well as their existing limited capacities to build resilience and adapt to adverse climatic impacts. The idea that global climate change magnifies and worsens existing development problems, in particular poverty, has been recognized and documented by the UN family of organizations and others for decades. An early consultative report prepared by a wide range of global development donors, including the African and Asian Development Banks, United Nations Development Programme (UNDP), United Nations Environment Programme (UNEP) and the World Bank, entitled *Poverty and Climate Change*, was presented during the 2002 UNFCCC's Conference of the Parties (COP) eighth session in New Delhi. The report recognized that there are variations as to the impacts of climate change and the vulnerability of poor communities to such change, but that generally climate change is superimposed on existing vulnerabilities and that, in many developing countries, climate change impacts are already development stressors (African Development Bank *et al.*, 2003).

Finding an effective resolution to global climate change has been highlighted repeatedly as a key global challenge by many leaders, including the UN

Secretary-General, who has identified climate change as a "defining issue of our time" since 2008 onwards (UN Secretary-General Statements, 2008).

In fact, the threat of climate change was made even more explicit in a 2014 US Department of Defense Report that was made publicly accessible, which referenced climate change as a "threat multiplier" and pointed out that "rising global temperatures, changing precipitation patterns, climbing sea levels, and more extreme weather events will intensify the challenges of global instability, hunger, poverty, and conflict" and "will likely lead to food and water shortages, pandemic disease, disputes over refugees and resources, and destruction by natural disasters in regions across the globe" (US Department of Defense, 2014, foreword). Some of the main impacts of global climate change that were emphasized in this US Department of Defense report include (p. 2):

- Reduction or melting of ice caps and glaciers
- Rises in ocean temperatures and acidity
- Increases in the frequency and intensity of extreme weather events (droughts, typhoons/hurricanes and super storms).

Figure 1.1 provides a schematic overview of the global processes and effects of climate change, which can be seen to have an impact on a wide range of human development activities that are relevant to all countries, and also distinguishes global warming and sea-level rise, which are separate yet related climate change processes.

More significantly, there has been mounting scientific evidence pointing to anthropogenic or human-induced global climate change that is simply impossible to discount or ignore. Two globally relevant scientific assessments have once again confirmed anthropogenic climate change as a key global challenge – the Fifth Assessment Report (AR5) of the Intergovernmental Panel on Climate Change (IPCC) and the Third US National Climate Assessment. The Summary for Policymakers (SPM) of Working Group I of the AR5 is emphatic in pointing out that the evidence for human influence on the global climate system has grown since its previous Fourth Assessment Report and that it is "extremely likely that human influence has been the dominant cause of the observed warming since the mid-20th century" (IPCC, 2013, p. 17). But it is the SPM's categorical caution regarding the irreversibility of a considerable portion of anthropogenic climate change in the absence of sustained action to curb CO_2 emissions that is stark:

A large fraction of anthropogenic climate change resulting from CO_2 emissions is irreversible on a multi-century to millennial time scale, except in the case of a large net removal of CO_2 from the atmosphere over a sustained period. Surface temperatures will remain approximately constant at elevated levels for many centuries after a complete cessation of net anthropogenic CO_2 emissions. Due to the long time scales of heat transfer from the ocean surface to depth, ocean warming will continue for centuries (IPCC, 2013, p. 28).

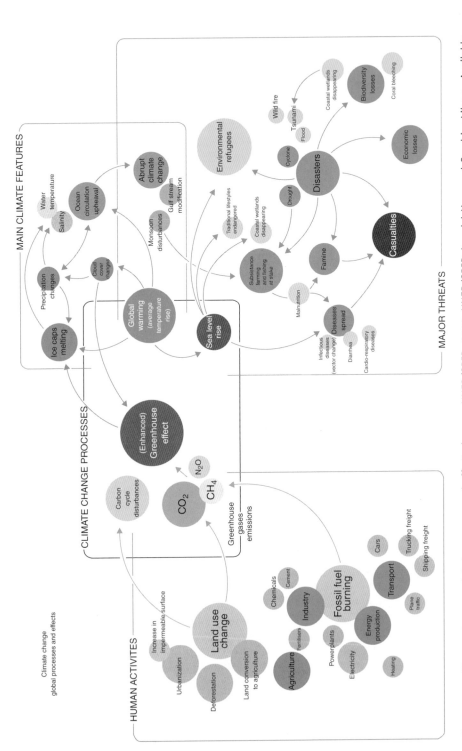

Figure 1.1 Climate change – global processes and effects (source: UNEP/GRID, 2009. UNEP/GRID-Arendal Maps and Graphics Library. Available at http://www.grida.no/graphicslib/detail/climate-change-global-processes-and-effects_9cc5#).

The grim IPCC finding about elevated surface temperatures being prolonged for centuries despite a complete stop of net anthropogenic CO_2 emissions not only underscores the importance of the precautionary principle in addressing climate change in a sustained manner, but also makes the inexorable pace in negotiating a globally relevant mitigation agreement harder to ignore. Box 1.1 contains key findings on the trends in stocks and flows of GHGs and their drivers as excerpted from the 2014 IPCC SPM of Working Group III.

Box 1.1 Key trends in stocks and flows of GHGs and their drivers

- Total anthropogenic GHG emissions have continued to increase over 1970–2010 with larger absolute decadal increases toward the end of this period (high confidence)
- CO_2 emissions from fossil fuel combustion and industrial processes contributed about 78% of the total GHG emission increase from 1970 to 2010, with a similar percentage contribution for the period 2000–2010 (high confidence).
- About half of cumulative anthropogenic CO_2 emissions between 1750 and 2010 have occurred in the last 40 years (high confidence).
- Annual anthropogenic GHG emissions have increased by 10 $GtCO_2e$ between 2000 and 2010, with this increase coming directly from energy supply (47%), industry (30%), transport (11%), and buildings (3%) sectors (medium confidence).
- Globally, economic and population growth continue to be the most important drivers of increases in CO_2 emissions from fossil fuel combustion. The contribution of population growth between 2000 and 2010 remained roughly identical to the previous three decades, while the contribution of economic growth has risen sharply (high confidence).
- Without additional efforts to reduce GHG emissions beyond those in place today, emissions growth is expected to persist, driven by growth in global population and economic activities. Baseline scenarios, those without additional mitigation, result in global mean surface temperature increases in 2100 from 3.7°C to 4.8°C compared with pre-industrial levels (median values; the range is 2.5–7.8°C when including climate uncertainty (high confidence).

Source: IPCC (2014).

In addition to the global scientific consensus presented in AR5, the recently released Third US National Climate Assessment is equally categorical about human-induced climate change, stating that:

> Natural drivers of climate cannot explain the recent observed warming. Over the last five decades, natural factors (solar forcing and volcanoes) alone would actually have led to a slight cooling. The majority of the warming at the global scale over the past 50 years can only be explained by the effects of human influences, especially the emissions from burning fossil fuels (coal, oil, and natural gas) and deforestation. The emissions from human influences that are affecting climate include heat-trapping gases such as CO_2, methane, and nitrous oxide, and particles such as black carbon (soot), which has a warming influence, and sulfates, which have an overall cooling influence (Walsh *et al.*, 2014, p. 23).

The National Climatic Data Center of the U.S. National Oceanic and Atmospheric Administration (NOAA) recently reported (as part of the services it provides to the public to support informed decision-making) that:

> 2014 was the warmest year across global land and ocean surfaces since records began in 1880. The annually-averaged temperature was 0.69°C (1.24°F) above the 20th century average of 13.9°C (57.0°F), easily breaking the previous records of 2005 and 2010 by 0.04°C (0.07°F). The 2014 global average ocean temperature was also record high, at 0.57°C (1.03°F) above the 20th century average of 16.1°C (60.9°F), breaking the previous records of 1998 and 2003 by 0.05°C (0.09°F) (NOAA, 2015).

Meanwhile, the State of the Climate 2013 report pointed out that, globally, 2013 was one of the 10 warmest years on record, both at the surface and in the troposphere, and that GHGs such as "carbon dioxide (CO_2), methane (CH_4), and nitrous oxide (N_2O) continued to increase in the atmosphere during 2013". The report pointed out that "on 9 May, for the first time since CO_2 measurements began in 1958 at Mauna Loa, Hawaii, the daily average mole fraction exceeded 400 parts per million (ppm)" (Blunden and Arndt, 2014, p. 5).

There is increasing evidence that extreme weather events are wreaking havoc in many countries across the world, and particularly in countries with low-lying and fragile coastal areas. In the past 5 years, the spate and intensity of extreme weather events – ranging from caustic heat waves in Russia, the catastrophic destruction in the Philippines caused by Typhoon Haiyan, and the ravaging floods in Pakistan, India, China, and the northeastern parts of the United States – have been seen as harbingers of regional global climatic changes (Cherian, 2012a). Figure 1.2, which is excerpted from NOAA's *2014 Annual Climate Report*, provides a schematic representation of selected significant climate anomalies and events in 2014.

By many global estimates, the risks of global climate change are not going away, and appear to be getting riskier and costlier to address (International

Selected significant climate anomalies and events in 2014

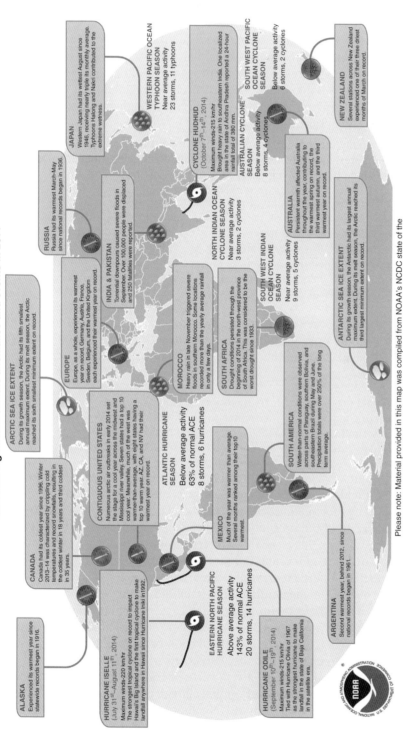

Figure 1.2 Climate anomalies and events in 2014 (source: NOAA: Global Analysis – Annual 2014 Climate Report. Available at http://www.ncdc. noaa.gov/sotc/service/global/extremes/201413.gif).

Energy Agency [IEA], 2011; World Economic Forum [WEF], 2014). Regardless of the uncertainties about the exact scope of climatic impacts on individual countries, global climate change has been seen as one of the definitive forces impacting current and future prospects for human development in all countries (World Bank, 2012). There is broad agreement that the costs of inaction in addressing global climate change outweigh the costs of action (Stern Review, 2006; IEA, 2009).

A recent WEF report clearly articulated the ripple or multiplier effects of the global community's delay in failing to act decisively in response to the risk of global climate change:

> The risk of climate change by far displays the strongest linkages and can be seen to be both a key economic risk in itself and a multiplier of other risks, such as extreme weather events and water and food crises... The risk of global governance failure, which lies at the heart of the risk map, is linked to the risk of climate change. Negotiations on climate change mitigation and adaptation are progressing by fits and starts, perpetually challenged to deliver a global legal framework (WEF, 2014, pp. 21–22).

The irony is that more than 20 years ago, the UNFCCC recognized through the adoption of the precautionary principle, the risk of waiting too long and doing too little to address the challenge of climate change. Contained in Article 3.3 is the explicit recognition that:

> Parties should take precautionary measures to anticipate, prevent or minimize the causes of climate change and mitigate its adverse effects. Where there are threats of serious or irreversible damage, lack of full scientific certainty should not be used as a reason for postponing such measures (IUCC, 1992, p. 6).

This prescient 1992 recognition calling on parties to take precautionary measures that address global climate change is often neglected in the recent global rush to call for bold action on climate change. The proverbial bottom line is that in spite of the precautionary principle being enshrined within the 1992 UNFCCC, over 20 years of arduous intergovernmental negotiations convened all over the globe, and mounting evidence compiled most recently by AR5 of the IPCC, the prognosis for securing a comprehensive consensus-based global climate change mitigation agreement by 2015 appears less than clear. The idea that bold action to mitigate against climate change is still urgently sought after more than 20 years of intergovernmental global climate change negotiations is sobering and provides a proximate rationale for the ideas outlined in this chapter and discussed further in the book.

Cognizant of the scale and scope of adverse effects of global climate change, the 1992 UNFCCC remains the only global climate change agreement that has near-universal ratification in terms of UN member states. But, after more than 20 years of climate change negotiations in cities all over the globe, there are still no immediate signs of a comprehensive global framework agreement to mitigate climate change, even though the 2015 deadline for such an agreement is fast approaching. The decades-long UN-led climate change negotiations have been looking like a quixotic quest for a comprehensive legally binding framework deal that is anticipated to loom over the negotiating horizon, but which is never quite resolved as negotiators travel annually from one global capital to another in search of a consensus-based agreement. But, after years of negotiating around the globe, the impending deadline of a final climate change agreement that will govern the post-2015 development agenda at the 21st session of the COP (COP-21) to the UNFCCC looms ever closer.

Addressing the challenge of climate change has been made enormously difficult because it requires countries to confront their energy needs and energy security concerns, which in turn impacts on their respective national socio-economic development plans. The linkages between lack of access to energy services and diminished opportunities for economic productivity have been recognized as mutually reinforcing components (Brew-Hammond, 2010). According to the IEA:

> Modern energy services are crucial to human well-being and to a country's economic development: and yet, globally 1.7 billion people are without access to electricity and 2.7 billion are without to clean cooking facilities. More than 95% of these people are either in sub-Saharan Africa or developing Asia, and 84% are in rural areas (IEA, 2011, p. 7).

The linkages between lack of access to modern energy services and income poverty, nutrition, gender inequality and human development have been referenced previously by a range of studies (Goldemberg *et al.*, 1988; Sokona *et al.*, 2004; Modi *et al.*, 2006). Access to modern energy services has been deemed crucial not only in enhancing socio-economic development but also in reducing poverty, and contributing to international security (Bazilian *et al.*, 2010; Sovacool, 2012).

These synergies between energy and socio-economic development, and the very real health and human welfare implications of consigning millions to depend on unreliable and heavily polluting traditional sources of energy can, and should be contrasted with the claim that there is "no inherent connection between the promotion of improved welfare for the poorest households and the reforming of energy markets" (United Nations Department of Economic and Social Affairs [UNDESA], 2007, p. 2), and the characterization of the "direction of causation" between the modernization of energy systems, economic growth and the improvement of human welfare as "less established" (UNDESA, 2007, p. 44).

In fact, it can be argued that it is the inherent linkages between energy and sustainable human development, including the nexus between energy, climate change and poverty reduction, that are crucial for any future UN-led post-2015 development agenda, because poverty eradication and climate change have long been highlighted as key areas in which UN-sponsored global agreements are being sought. It is precisely the primacy accorded to the role of energy in fueling both socio-economic development and anthropogenic climate change that translates to make the global goal of mitigating energy sector emissions such a thorny political issue for the world's largest aggregate GHG emitters. Despite a broad global consensus that anthropogenic emissions of CO_2 and other GHGs are the leading cause of climatic changes (World Bank, 2012; IPCC, 2013), progress to mitigate climate change through a comprehensive framework agreement that would limit GHG emissions has been both very slow and difficult. Furthermore, it has been estimated that GHG emissions from developing countries are likely to exceed those from developed countries within the first half of this century (IEA, 2009, 2011), which complicates the ongoing global discussion on allocating responsibilities for addressing anthropogenic global climate change in the lead-up to COP-21 in 2015. But the problem of climate change is more challenging for developing countries that have simultaneously to deal with increasing energy access for their poor, whilst also addressing energy-related mitigation. In other words, there is another, somewhat less discussed, aspect of the energy and climate change relationship, which is that close to 1.3 billion people lack access to sustainable and cost-effective energy services, and 2.7 billion rely heavily on polluting and unsustainable forms of energy to secure their basic human needs.

The number of poor on a regional basis who lack access to modern energy services are hard to ignore, especially in light of the UN-led quest for a sustainable future for all. In its 2011 report outlining a review of good practices and lessons learned in terms of energy access for the poor in the Asia-Pacific region, the UNDP noted that "nearly half of the world's population still lacks reliable access to modern energy services". According to the report: "Roughly 2.7 billion people (40 percent of the world's population) depend on traditional biomass for cooking and 1.4 billion remain without access to electricity", but there is a regional thrust to this problem because in the Asia-Pacific region "almost 2 billion people are dependent on the traditional use of biomass", and almost 800 million in this region have no access to electricity (UNDP, 2011, p. 15).

In reviewing the energy access situation in developing countries, UNDP/World Health Organization (UNDP/WHO) notes that 44% of those who die each year from household air pollution resulting from the use of traditional biomass and ineffective energy services are children, while women account for 60% of all adult deaths (UNDP/WHO, 2009). In 2008, the WHO pointed out that the number of people estimated to die every year due to household air pollution from poorly combusted biomass fuels is expected to rise by 2030 to around 1.5 million (WHO, 2008). More recently, the WHO has outlined the linkages between indoor air pollution and household energy use for "the forgotten 3 billion" poor people who rely heavily on solid fuels and inefficient cooking fuels and technologies that produce high levels of household air pollution, including small soot particles that penetrate deeply into lungs. According to the WHO, indoor smoke can be 100 times higher than acceptable levels for small particles in poor households relying on inefficient energy sources and systems, risks of exposure are particularly high among women and young children, and this heavy reliance on solid fuels for cooking and household needs by the poor has resulted in over 4.3 million premature deaths in 2010 due to indoor air pollution (WHO Media Centre, 2014).

The lack of access to modern energy services has been termed "energy poverty" and it is considered to have lasting impacts on health, education, and employment for the poor. The energy-poor, for instance, suffer the health consequences of indoor air pollution due to the inefficient combustion of solid fuels and biomass, as well as the economic impacts of their lack of access to energy for sustaining livelihoods and education opportunities (UNDP, 2011, p. 13). Currently, many poor communities, primarily in developing countries located in Asia and Africa, lack access to safe and reliable energy, and pay disproportionately high prices for inefficient and ill-health-generating energy sources. In many developing countries, poor households spend more than a third of their household expenses on poor-quality sources of energy; and poor women and young girls spend a disproportionate amount of daily time and effort in securing inefficient solid fuels (traditional biomass), which gives them less time for productive employment/livelihood opportunities (Modi et al., 2006).

Table 1.1, which has been excerpted from Organisation for Economic Co-operation and Development (OECD)/IEA 2011 special report, *Energy For All: Financing Access for the Poor*, provides an overview of the numbers and regional nature of representation of people without access to modern energy services, including those without access to electricity and those who rely heavily on the traditional use of biomass for cooking.

While the broad topic of "energy access for the poor" is considered at length in this book, it is important to note at the outset that the concept of energy access has proven to be challenging to define, categorize, and implement. Recognizing that there was no universally agreed and adopted definition of energy access, the IEA sought to provide a clearer definition of energy access as: "a household having reliable and affordable access to clean cooking facilities, a first connection to electricity and then an increasing level of electricity consumption over time to reach the regional average" (IEA, 2011, p. 12). Pointing to the 20-year trajectory of analysis focused on the topic of access to modern energy services, and its role in poverty reduction and economic development, Sokona et al. (2012) have argued that responding to the challenge

Table 1.1 Number and share of people without access to modern energy services in selected countries, 2009

	Without access to electrcity		Relying on the traditional use of biomass for cooking	
	Population (millions)	Share of population	Population (millions)	Share of population
Africa	587	58%	657	65%
Nigeria	76	49%	104	67%
Ethiopia	69	83%	77	93%
DR of Congo	59	89%	62	94%
Tanzania	38	86%	41	94%
Kenya	33	84%	33	83%
Other sub-Saharan Africa	310	68%	335	74%
North Africa	2	1%	4	3%
Developing Asia	675	19%	1,921	54%
India	289	25%	836	72%
Bangladesh	96	59%	143	88%
Indonesia	82	36%	124	54%
Pakistan	64	38%	122	72%
Myanmar	44	87%	48	95%
Rest of developing Asia	102	6%	648	36%
Latin America	31	7%	85	19%
Middle East	21	11%	0	0%
Developing countries	1,314	25%	2,662	51%
World*	1,317	19%	2,662	39%

*World total includes Organisation for Economic Co-operation and Development (OECD) countries and eastern Europe/Eurasia.
Source: IEA (2011, p. 11). Available at http://www.iea.org/publications/freepublications/publication/weo2011_energy_for_all.pdf

of increasing access of energy services requires a cross-sectoral approach to development planning, and requires the "energy access debate" to be "broadened towards a system-wide treatment of the energy issue in development" (p. 4). According to Sokona *et al.* (2012), it is the idea of expanding the concept of energy access and thereby addressing energy access in the context of the broader energy transition that makes it "imperative that countries need to develop their future energy systems with due consideration to climate change issues" (p. 5). From the perspective of this chapter and this book, it is exactly this expansion of the concept of energy access for the poor in terms of its linkages with climate change and sustainable development that is the main focus of investigation. In this regard, the argument that is advanced is not whether and how to increase energy access for the poor, or whether and how to improve the prospects for a comprehensive climate change agreement, but instead to ask whether and how energy access for the poor has been factored into the series of key UN-led intergovernmental climate change and sustainable development outputs.

At the intergovernmental level, the vital role of energy in driving human development was explicitly recognized as early as the historic 1972 Stockholm Conference and the 1983 World Commission on Environment and Development. Goldemberg *et al.* (1985, 1988) provided an early understanding and recognition of the links between energy and a host of other development concerns, including income poverty, malnutrition, and ill-health. Evidence underscoring these linkages between energy and poverty, gender and health in terms of the impacts of household energy use on indoor air pollution and health of women and young children was raised early by the WHO (2000). This lack of access to sustainable and cost-effective energy services for millions of people who live primarily in South Asia and sub-Saharan Africa is a different aspect of the climate change and sustainable development relationship that has to be reconciled within the UN's quest to secure a shared post-2015 development agenda.

The multidimensional development linkages among energy and poverty, women, urbanization, and production and consumption patterns were documented in a 2000 report jointly prepared by two of the main development agencies of the UN, and the World Energy Congress (WEC). The report noted that although current energy consumptions patters are unsustainable, energy can help to solve global concerns, particularly those related to poverty, gender, and urbanization, amongst others. But according to the report, "to realise this potential, energy must be brought to centre stage and given the same importance as the other major global issues" (UNDP/UNDESA/WEC, 2000, p. 40).

Over the past two decades, poverty reduction and global climate change have been identified as two of the most pressing challenges in the UN's quest for a shared development agenda. Energy has clearly been identified as a key driver in both global challenges, but the question is whether there are concrete references and recommendations on how to address energy access for the poor in key globally agreed outputs emanating from the global climate change and sustainable development negotiations. This chapter will provide a framework for the argument that the linkages between climate change and energy access for the poor are crucially important to UN's quest for shared post-2015 sustainable development agenda precisely because energy access for the poor cuts across two of the most urgent development-related challenges: climate change and poverty reduction. The well recognized linkage of modern energy access to development and environmental objectives is what make the concept of energy access a "central element of the debate on sustainable development" (Rehman *et al.*, 2012, p. 27). Perhaps the simplest and starkest way to understand the human costs of this nexus is that the heavy reliance on inefficient energy sources contributes to emissions of short-lived climate pollutants (SLCPs) that impact negatively on the daily lives and health of millions of poor people. And it is this neglected nexus between energy access for the poor and climate change mitigation that has not received adequate global attention to date. As the evidence presented in this book will demonstrate, the linkages between climate change and poverty reduction, especially in terms of increasing energy access for the poor, have not been referenced in a systematic and organized manner within the context of negotiated and agreed global climate change outcomes.

Coincidentally, at the very time that UN climate change negotiations appear to be progressing fitfully, there has been a push by the UN Secretary-General to advocate for linkages between energy and climate change under the aegis of the Advisory Group on Energy and Climate Change (AGECC), established in 2010. The AGECC was convened to address the dual challenges of meeting the world's energy needs for development while contributing to a reduction in GHGs, and it resulted in calling on the UN and its member states to commit themselves to two key goals (UN Secretary-General's AGECC, 2010, p. 9):

- Ensure universal access to modern energy services by 2030.
- Reduce global energy intensity by 40 percent by 2030.

In 2011, AGECC was subsumed by the UN Secretary-General's initiative known as the Sustainable Energy for All (SE4All). Currently, led by the UN and the World Bank, SE4All includes an additional goal of doubling the share of renewable energy added to the twin goals identified previously. However, the intergovernmental response to the advocacy of these goals has been mixed and the lackluster reference to the SE4All in the 2012 Rio+20 Conference is a factor that is discussed later in the book.

In an urgent bid to scale up high-level climate change action prior to the globally agreed 2015 deadline by which intergovernmental negotiators are supposed to secure a consensus-based comprehensive climate agreement, the UN Secretary-General convened a special high-level Summit on Climate Change on September 23, 2014, where heads of state and leaders of business and civil society were invited to come with "bold and new announcements and action" (UN Secretary-General Statements, 2013). Reiterating that "climate change is the defining issue of our time", at a preparatory meeting held in Abu Dhabi (May 4, 2014) ahead of the 2014 Summit, the Secretary-General called on governments to "complete a meaningful new climate agreement by 2015 that will rapidly reduce emissions and support resilience" and "identified nine key areas with the greatest potential for fast, mean-ingful results" which include "energy, cities and transport, finance, resilience, agriculture and short-lived climate pollutants"(UN Secretary-General Statements, 2014).

With the leaders of the top aggregate GHG emitter, China, and the third largest aggregate emitter, India, not in attendance, along with the leaders of key countries such as Australia, the Russian Federation, and Germany, news reports were quick to dampen expectations of bold and new announcements emerging. Bloomberg reported that the leaders of China and India skipping the Climate Summit can be seen as signaling their "tepid support for a global pact to cut greenhouse gases among two of the largest emitters" (Yoon and Drajem, 2014). But there can be little doubt that the role of sustainable energy services. including technologies and systems, was highlighted by many heads of state and government from developing countries at the 2014 Climate Summit. In fact, Zhang Gaoli, the Special Envoy of the Chinese President, was quite clear about the predominant role that the energy sector would play in addressing climate change stating that: "China will advance a revolution in energy production and consumption, cap total energy consumption, raise energy efficiency and vigorously develop non-fossil fuels". Gaoli went on to emphasize the role of cooperation amongst developing countries – South-South cooperation – stating that "China will work hard to promote South-South cooperation on climate change"; and that China would double its annual financial support for the establishment of the South-South Cooperation Fund on Climate Change and provide an additional "six million US dollars to support the UN Secretary-General in advancing South-South cooperation on climate change" (UN Climate Summit 2014/Executive Office of the UN Secretary-General, 2014, p. 2).

The role of South-South cooperation in addressing climate change will, in all likelihood, grow in significance, given that China and India are among the largest aggregate GHG emitters. The importance of energy as a driver in economic development and anthropogenic climate change is one that has been recognized by countries like China and India, but not necessarily reflected in terms of programmatic guidance in the actual globally agreed climate change outcomes. The neglected part of the climate change–energy nexus, namely the linkages between climate change and increasing access to sustainable energy services for the poor, has received very little by way of concrete global programmatic focus and scaled-up action to date. It is precisely this neglected nexus between energy access for the poor and climate change that has not been adequately referenced in past global outcomes related to climate change and sustainable development that is the focus of this book.

The initial identification of energy and SLCPs as two of nine key action areas for climate action at the 2014 Climate Summit can be seen as significant and long overdue, because these sectors represent the potential for addressing the often-neglected nexus between climate change and energy access for the poor, but the subsequent absence of SLCPs as an action area in the final Summit outcome acts as a potential constraint. In the end, the action area of SLCPs was not included in the list of eight final action areas identified in the

2014 Climate Summit outcome, and a concrete opportunity to focus on reducing SLCPS in terms of energy access for the poor was perhaps missed. However, the Summit's summary outcome on energy does note that the SE4All initiative has set 2030 as a goal for doubling the global rate of energy efficiency improvement, doubling renewable energy's share in the global energy mix, and ensuring universal access to modern energy services (UN, Action Areas/Summit Announcements, 2014).

The reality is that the 2014 Climate Summit hosted by the UN Secretary-General was not the first time global attention has been focused on this issue. A high-level event focused on climate change was held on September 24, 2007 in the UN, entitled "The Future is in our Hands: Addressing the Leadership Challenge of Climate Change", and the first-ever thematic plenary debate of the UNGA, also held in 2007, focused solely on climate change and was entitled "Climate Change as a Global Challenge". In fact, 2007 was a watershed year in terms of focusing global attention on climate change with the nomination and subsequent joint awarding of the Nobel Peace Prize to Al Gore and the IPCC for their respective work on climate change (Gore, 2006). A few days after the UN special event on September 28, 2007, President Bush sponsored a conference meeting of the 16 "major emitters" of GHGs, namely Australia, Britain, Brazil, Canada, China, France, Germany, India, Indonesia, Italy, Japan, South Korea, Mexico, Russia, South Africa, and the United States, which together account for more than 90% of global GHG emissions. The stated objective of the meeting was to initiate a process by which the world's biggest emitters will outline targets for reducing their emissions, including outlining national measures for curbing emissions, setting long-term pollution objectives and seeing how smart technology, forestation, and financing for developing countries can help the carbon clean-up. This grouping of "major emitters" has been discussed as a forum for putting together voluntary measures to address global warming as opposed to comprehensive and mandatory targets and limits on emissions, and as a mechanism for getting the world's largest developing and developed country emitters together for the first time outside of the official UNFCCC processes (Johnson, 2007). The idea of a major emitters group was more formally launched by President Obama as the Major Economies Forum (MEF) and is discussed later in the book, as it continues to play a role in terms of bringing together key emitters, including key developing countries such as Brazil, South Africa, India, and China (also referred to as the BASIC countries).

A little over 7 years later, it is possible to argue that the collective will to secure a binding consensus-based climate mitigation framework agreement has clearly not been due to a paucity of global fora/venues for intergovernmental meetings and negotiations. Climate change negotiators have traversed the globe seeking the elusive consensus deal, but with leaders of key countries like China, India, Australia, and Germany absent from the 2014 Summit, persisting concerns about the global political will to secure a binding agreement by COP-21 in 2015 exist.

Twenty-two years after the UNFCCC agreement, the 20th COP (COP-20) in Peru resulted in the Lima Call to Climate Action (LCCA) with its "commitment to reaching an ambitious agreement in 2015 that reflects the principle of common but differentiated responsibilities and respective capabilities, in light of different national circumstances" (UNFCCC, 2015, p. 2). Key distinguishing features of the Lima COP-20 outcome are the "invitation to each Party to communicate to the secretariat its intended nationally determined contribution towards achieving the objective of the Convention as set out in its Article 2", and the agreement that each Party's intended nationally determined contribution aimed at achieving Article 2 – the UNFCCC's objective – "will represent a progression beyond the current undertaking of that Party" contained in paragraphs 9 and 10 (UNFCCC, 2015, p. 3).

These "intended nationally determined contributions" (INDCs) can therefore be anticipated to be the key component of any proposed new agreement that might be agreed at COP-21 in December 2015. INDCs can be characterized as a historic, first step towards a more inclusive coverage of countries, in contrast to the 1992 UNFCCC's, and the 1997 Kyoto Protocol's, references to Annex I Parties' (developed countries) obligations as being distinct from the roles and needs of non-Annex I Parties (developing countries). In paragraph 11, Parties are invited to "communicate" their INDCs "well in advance of" COP-21 (UNFCCC, 2015, p. 3), but paragraphs 7 and 8 of the Lima decision, which precede the reference to INDCs, are categorical in pointing out that any negotiating text or any submission of INDCs is "without prejudice to the legal nature and content of the intended nationally determined contributions of Parties or to the content of the protocol, another legal instrument or agreement" that might be agreed to at the Paris COP in 2015 (UNFCCC, 2015, p. 3). In fact, there is no formalized template or exact guidance as to the full scope of the quantifiable information being communicated, and/or agreement specifying any review or assessment processes that can be used for the reporting of these INDCs. INDCs submitted by a Party are supposed to represent a progression beyond the current undertaking of that Party and the question is how exactly national "progression" can be accounted. Paragraph 16 of the LCCA merely calls on the UNFCC Secretariat to publish on its website the INDCs as communicated by Parties; and to "prepare a synthesis report of aggregate effect of the intended nationally determined contributions communicated by Parties by 1 October 2015" (UNFCCC, 2015, p. 3).

In terms of the Lima COP outcomes, it is actually the Annex to the LCCA, entitled "Elements for a draft negotiating text", that is most significant, because it is this Annex that can be seen as the negotiating blueprint of sorts for the upcoming Paris COP-21. The Annex consists of a listing of various options and proposals put forward by different countries and groups of countries and provides an important footnote reminder that the "elements for a draft nego-tiating text reflect work in progress" and these elements "neither indicate convergence on the proposals presented nor do they preclude new proposals from emerging in the course of the negotiations in 2015" (UNFCCC, 2015, p. 6) But, as Davenport (2015) notes, the Lima outcome does not include any legally binding commitments that specify any particular amount of emissions reductions that countries need to cut; instead countries are requested to submit national INDCs by March 31 and those that miss this deadline are supposed to submit their INDC plans by June 2015. As of April 22, 2015, the UNFCCC Secretariat website lists that INDCs were communicated by only the following Parties: Switzerland, Latvia and the European Commission on behalf of the European Union and its member states, Norway, Mexico, the USA, Gabon, and Russia (UNFCCC/INDC website, 2015).

Interestingly, the LCCA and its Annex does not reference the terms "energy access" or "access to energy" but it does contain references in relation to paragraph 53.1 on "private and alternative finance" that call for, amongst other things, "a tax on oil exports from developing to developed countries to established", "an international renewable energy and energy efficiency bond facility to be established", and "the phasing down of high-carbon investments and fossil fuel subsidies" (UNFCCC, 2015, p. 24). Each of these three proposals has serious implications for the future of global energy scenarios and climate change mitigation, and when taken together can signal massive changes for sustainable development at the national and global level, so it will be impor-tant to see whether and how they factor into the final agreed text at the upcoming Paris COP in 2015. These proposals for a tax on oil exports from developing countries to developed countries, as well as the phasing down of high carbon investments and fossil fuel subsidies, have the potential to completely alter not just the global energy landscape, but also the financial and development arenas, particularly for developing countries that are heavily dependent on fossil fuels for socio-economic growth. This proposal to estab-lish a tax on oil exports from developing to developed countries is also puzzling in terms of its placement within the Annex, precisely because the idea for a tax on oil exports from developing to developed countries has no concrete referential analog in any of the key agreed global decisions of the over 20-year UNFCCC process, starting from the UNFCC itself to the 2013 Warsaw COP-19 outcomes. A search of all the key agreed outcomes, from the UNFCCC to the Kyoto Protocol, the Berlin Mandate and on to the 2013 Warsaw COP-19, indicates that no proposal on establishing a tax on oil exports from developing to developed countries has ever been agreed to before. The establishment of tax on oil exports and the phasing out of high carbon investments and fossil fuel subsidies have massive implications for future global financing and sustainable development, but happen to be listed in a sub-section dealing with "private and alternative finances". What is also interesting is that the draft proposal suggests that only oil exports from developing countries to developed countries would be targeted for taxation, which could impose new responsibilities and costs associated only with oil exports from developing countries, and which can be seen as a new means of trade and environmental conditionality on oil exports of developing countries.

In the lead-up to the Paris COP-21, the draft negotiating text is continually growing in length and subject to change. The most current version of the draft negotiating text emanating from the post-Lima negotiations process – the February 2015 session of the Ad Hoc Group of the Durban Platform – indicates that the proposal for a tax on oil exports is contained in paragraph 128.1 and now reads: "A tax on oil exports from [developing] [Parties not included in annex X] to [developed countries] [Parties included in annex X][Parties in a position to do so, considering evolving capabilities] [all countries in a position to do so] to be established" (UNFCCC/ADP, 2015, p. 50). The brackets around the terms "developing" and "developed" and the introduction of newly introduced and additional bracketed references indicate that the proposal, like many other elements of the negotiating text, is neither agreed to nor without contestation. The other two proposals, namely the establishment of an international renewable energy and energy efficiency bond facility and the phasing down of high carbon investments and fossil fuel subsidies, currently remain without brackets, which indicates that global climate change negotiators have not suggested or proposed changes to these two proposals to date. While the Annex to the LCCA did not contain any references on energy access, the most recent version of the draft negotiating test does now contain one solitary option referencing "universal energy access", which is contained in paragraph 176.1 of the section dealing with time frames/commitments, which states: "Each Party to consider adjustments on the basis of historical responsibilities and equitable sharing of global atmospheric resources and carbon space in the context of imperatives of poverty eradication, universal energy access and sustainable development for developing countries" (UNFCCC/ADP, 2015, p. 75). However, it remains to be see whether and how these proposals, which have serious implications for future global and national energy and sustainable development arenas, will fare in terms of the final agreement at the end of COP-21.

In light of these proposals, and given the absence of concrete, specific guidance linking climate change and energy access for the poor within the LCCA and in the current draft negotiating text, the overall question that needs to be asked is whether the nexus between climate change energy-related mitigation and energy access for the poor will ever be explicitly reflected in any comprehensive global climate change agreement arrived at in 2015. The year 2015 is shaping up to be a pivotal one for agreement on the UN's post-2015 development agenda, anchored by an ambitious set of sustainable development goals, including separate stand-alone goals on climate change and energy access. It stands to reason that increasing access to clean and affordable energy services for the poor should serve as a crucial link between two separate sets of UN negotiations on climate change and sustainable development, which are both expected to yield global agreements in 2015. In other words, linkages among energy access, poverty reduction and climate change cannot and should not be ignored by the UN quest for a comprehensive global climate change agreement and a shared post-2015 development agenda.

What this book will demonstrate is that the heavy reliance on inefficient and polluting energy services by the poor, including, more specifically, the challenge of worsening air pollution due to emission of SLCPs and attendant disease burden resulting from a lack of access to sustainable energy services, has not been adequately focused despite decades of global negotiations on climate change mitigation and sustainable development. Energy access for the poor has not been referenced in key agreed global outputs resulting from both sets of negotiations. Despite years of global negotiations, an examination of the actual record of key UN globally agreed climate change and/or sustainable development-related outcomes reveals a surprising paucity of programmatic guidance and references regarding the linkages between lack of access to sustainable energy services for the poor and climate change.

1.2 Framing the argument and issuing the necessary caveats

This chapter and the ones that follow will not provide a detailed accounting of all UN-led climate change and sustainable development negotiations, nor will there be any detailed discussion on the specific roles that key countries, actors, and institutions play in these negotiations. Background information on the climate change negotiations can be found from a variety of sources, including the UNFCCC official website maintained by the UNFCCC Secretariat. A detailed assessment of different options and approaches that address various aspects of the post-2012 international climate change negotiations processes and associated mechanisms can be found in a joint UNEP/World Resource Institute (WRI) paper that reviews five key issues which are seen as critical to the international climate change regime: options under the UNFCCC to increase ambition beyond existing commitments and action; options outside the UNFCCC to increase ambition beyond existing mitigation commitments and actions; means for sharing the mitigation effort under the UNCCC; the role of various actions in tracking country performance on mitigation; and the legal form of a future climate agreement (UNEP/WRI, 2011).

Instead, the focus is to provide a framework for the argument that there is a puzzling disconnect between energy access for the poor and climate change within the parallel track process of UN-negotiated global outcomes on climate change and energy. This chapter argues that energy issues are critical to any UN-led comprehensive global climate change agreement, and that increasing access to sustainable energy services for the poor is equally central to any UN long-term global sustainable development agenda predicated on poverty eradication. It would stand to reason that the nexus between energy access for the poor and climate change would be crucial in terms of achieving climate change, poverty eradication, and sustainable development objectives. The central premise explored is whether the nexus between energy access for the poor and climate change has been focused on, or has been neglected within the context of key UN agreed outcomes on climate change and sustainable development. The question raised, therefore, is not why or who is responsible for the fact that a consensus-based global climate change mitigation agreement has not been reached so far, but rather whether the issue of energy access for the poor has been adequately referenced in terms of agreed global outcomes on climate change and sustainable development over the course of many decades. The focus is on understanding what the agreed record of more than 20 years of global negotiations on climate change and 40 years of global sustainable development conferences reveals in terms of concrete policy and programmatic guidance and inputs on the linkages between energy access for the poor and climate change.

To be clear, the arguments presented do not purport to provide a detailed timeline or exact historical accounting of the 20-year process of climate change negotiations or the 40-year process of sustainable development negotiations. Instead, the main concern is to examine key agreed global outcomes of the past 20 years of UN-led intergovernmental negotiations on global climate change, and 40 years of sustainable development in terms of their concrete references and guidance on energy access for the poor. The focus is carefully examine key UN outcomes on climate change and sustainable development with a view to providing concrete textual analysis as to whether and how energy access for the poor has been referenced within two sets of global negotiation tracks ranging from the 1992 UNFCC to the 2014 LCCA, and from the 1972 Stockholm Conference to the 2012 Rio+20 Conference and its follow-up.

It should be clearly noted that the arguments presented here will not make a case for analyzing the causes of, or assigning responsibilities for, the expunging of "energy" and/or "energy access for the poor" from the intergovernmental global climate change negotiations. Additionally, the arguments presented will not analyze past or existing energy-related mitigation projects, including those submitted to the Clean Development Mechanism, nor will we provide any kind of statistical analyses of nationally reported emissions trends related to the UNFCCC process or weigh in with policy options related to emissions trading or carbon caps.

It is also extremely important to point out that while referencing the central role of climate change adaptation, and underscoring that need for adaptation is hugely urgent in developing countries and especially so in poor communities, the book will not focus on or discuss adaptation efforts or adaption in general. The role of adaptation is therefore taken as vitally important in terms of addressing climate change and sustainable development in general. The need to link ongoing adaptation efforts with broader national sustainable development objectives, including poverty reduction, should be seen as extremely urgent and effective, particularly for developing countries where resources (financial and capacity) are limited. The need to focus on the linkages between building resilience and reducing vulnerabilities of the poor to adverse impacts of climate change-related disasters has been touched upon in a special report by the IPCC on managing the risks of extreme events and disasters to advance adaptation (IPCC, 2012).

Additionally, while highlighting the importance of climate financing and capacity building as crucial means of implementation for delivering on climate change-related objectives, it is important to clarify that neither climate financing nor capacity building are the immediate focus of the arguments advanced in this book. Insights into the current or past state of climate finance and climate funds, and the various funds associated with the capacity building within the UNFCCC process and outside of it are, however, highly relevant and extremely urgent. The issue of climate finance has been the subject of considerable debate in intergovernmental fora, but exact assessments of the scope of climate finance still remain unclear, because currently there is no globally recognized framework accounting for cumulative climate financing that comprehensively unifies both public and private sector flows. As the 2014 IPCC SPM Working Group III report noted that there "is no widely agreed definition of what constitutes climate finance" and although estimates are available, the IPCC indicates only medium confidence in its statement that "published assessment of all current annual financial flows whose expected effect is to reduce net GHG emissions and/or enhance resilience to climate change and climate variability show USD 343 to 385 billion per year globally" and of *"this total, public climate finance that flowed to developing countries is estimated to be between USD 35 to 49 billion/yr in 2011 and 2012"* (IPCC, 2014, p. 29). These best-case estimates for cumulative (mitigation and adaptation) and developing country-related climate funding are a far cry below what may actually be needed for just adaptation with studies that suggest that developing countries' needs for adaptation actions may be in the range of US$100–450 billion a year (Montes, 2012).

From the perspective of the intergovernmental climate change process, it is clear that the Green Climate Fund (GCF), which was established at COP-16

(Cancun, 2012) as an operating entity of the financial mechanism of the UNFCCC (in accordance with Article 11), has broad aspirational goals. The GCF's stated mandate is to contribute to the overall objective of the Convention and sponsor the proverbial paradigm shift:

> In the context of sustainable development, the Fund will promote the paradigm shift towards low-emission and climate-resilient development pathways by providing support to developing countries to limit or reduce their greenhouse gas emissions and to adapt to the impacts of climate change, taking into account the needs of those developing countries particularly vulnerable to the adverse effects of climate change (GCF website).

In the meantime, COP-20 in Lima acknowledged that the GCF had mobilized US$10.2 billion to date from contributing Parties, "making it the largest dedicated climate fund" (UNFCCC website, 2015). Concerns remain over transparency, procedures, and the role of civil society participation in the GCF, as well as worries that private sector investors will not steer funds away from adaptation, which is of "secondary interest to investors who have a clearer route to profits from projects that reduce emissions" (Parnell, 2013). As May 2015, the GCF has finally obtained the requisite 50 percent threshold of signed contributions and can begin disbursing funds to developing countries.

Rather than speculating on the scope of climate finance, the scope of analysis is circumscribed to examining the agreed record of UN-led negotiations on both climate change and sustainable development. The actual global outcomes of UN negotiations on climate change will be carefully studied, including the UNFCCC, the Kyoto Protocol, the Bali Road Map, the Copenhagen Accord, the Cancun Agreement, the Durban Outcomes, the Doha Climate Gateway, the Warsaw Outcomes and the Lima Call to Action, as well as a series of historic UN conferences and summits on environment and sustainable development, ranging from the 1972 United Nations Conference on Human Environment (UNCHE) to the 1992 UNCED, the WSSD and the 2012 Rio+20 Summit, including the Millennium Development Goals (MDGs) and the energy-related negotiations within the UN's Commission on Sustainable Development.

It is these actual agreed global outcomes of UN processes that are being examined, rather than the mechanics, norms, groups of individuals or institutions related to the negotiations processes, and it is these agreed outcomes on climate change and sustainable development that serve as the principal primary source material in order to see how and whether the energy access–poverty reduction–climate change nexus has been referenced over the course of several decades.

What makes the evidence presented in this book unique is that the agreed outcomes of two sets of UN-led intergovernmental negotiations, namely climate change and energy for sustainable development, have not been previously examined in terms of the substantive and concrete linkages on energy access for the poor. The actual record of key outcomes emanating from the two different and separate UN-led intergovernmental negotiating tracks on climate change and sustainable development have not been accessed in depth in terms of energy access for the poor objectives.

The idea that two pressing and linked development challenges facing the UN global community have been dealt with as separate and largely disconnected negotiating tracks requires careful consideration. But even before discussing the exact content and scope of the key UN global outcomes on climate change and energy, it is important to focus attention on key elements and factors that inform the energy access–poverty reduction–climate change nexus, because these elements provide a necessary frame of reference for the UN-led negotiations on these issues. In other words, why does energy access matter for poverty reduction and impact on global climate change?

The aim of Chapter 1 is therefore:

- to outline the escalating risks of global climate change;
- to provide a better understanding of the nexus between climate change and energy access for the poor in developing countries.

This chapter establishes the framework for the argument that climate change and energy access for the poor cannot be viewed within separate silos, because energy is a key driver in addressing both climate change and sustainable development. The policy and programmatic nexus between these challenges is crucial for their urgent and equitable resolution in terms of the broader UN-led post-2015 development agenda.

Existing analyses related to the linkages between the role of energy access for poor and climate change will be provided to argue that addressing the energy–poverty–climate change nexus is highly relevant to the ongoing work of the UN and the global development community and the UN-led negotiations on climate change and sustainable development. The final section of this chapter outlines the idea that having two separate negotiating tracks is not conducive to promoting linkages amongst these global concerns, especially in light of the current UN search for an integrated post-2015 development agenda. The absence of concrete guidance and references linking these three concerns constitutes both a huge missed opportunity and also a larger concern for the UN's current intergovernmental negotiations, which seek to move the UN and its member states towards a shared and universal post-2015 development agenda.

Chapters 2 and 3 focus on the main question of whether the nexus between energy access for the poor and climate change has been adequately referenced in the parallel yet separate UN-led negotiations on climate change and sustainable development. The bulk of the book, namely

Chapters 2–4, focus on what key intergovernmental outcomes on climate change and sustainable development over the course of multiple decades actually state and reference in terms of the nexus between energy access for the poor and climate change. The objective is to provide a detailed examination of a series of key UN globally agreed climate change outcomes and outline the exact references made in these outcomes on energy access for the poor.

Chapter 2 examines what the record of key outcomes of two decades of global climate change negotiations reveals in terms of concrete guidance and references related to energy access for the poor. Having separate and parallel track negotiations for global climate change and sustainable development has been justified on the basis that the negotiations under the aegis of the UNFCCC are legally binding to the parties who have signed and cannot be therefore integrated into the broader UN process on sustainable development. But what is harder to reconcile is why energy issues in general, and concrete programmatic inputs on energy-related mitigation are largely absent from key global climate change outcomes over the course of more than two decades.

Chapter 3 examines the record of close to 40 years of UN global conferences on environment and development and sustainable development, which is where energy and energy for sustainable development issues have been discussed, to see what specific references and linkages have been made to climate change, energy access, and poverty reduction. Taken together, Chapters 2 and 3 review evidence of decades of UN-led global outcomes on climate change and sustainable development, which demonstrate the overwhelming trend to separate negotiating silos on energy for sustainable development and climate change with little to show in terms of integrated action on increasing energy access for the poor. The long history of separate and parallel negotiated outcomes on energy for sustainable development and global climate change will be highlighted.

Chapter 4 continues the focus of examining key global outcomes on sustainable development, but this time the object is to look at how the role of non-legally binding voluntary initiatives such as partnerships and voluntary commitments have been referenced, and in particular how climate change and energy access for the poor factor into these voluntary activities. Accordingly, Chapter 4 looks at the rapid growth and diverse array of voluntary initiatives emerging from the 1992 Agenda 21 to the recent 2012 Rio + 20 Summit focused on a variety of sustainable development concerns and argues that these voluntary initiatives offer interesting and innovative options for future action, but also raise serious concerns in terms of lack of institutional clarity as to how to differentiate amongst different initiatives, and the lack of an integrated accountability framework by which these voluntary initiatives can be comprehensively accessed and evaluated.

Moreover, the self-reporting nature of voluntary efforts related, for instance, to sustainable energy and climate change makes it hard to evaluate and differentiate amongst a plethora of voluntary activities, and the overall lack of a clear, universal, globally accessible accountability framework for such activities raises accountability and transparency concerns for any future UN-led post-2015 development agenda, with its universal agenda and shared goals.

The final chapter of the book provides concluding evidence for the argument that having separate, parallel negotiating silos on climate change and the energy-related needs of poverty reduction within the UN has been and remains a waste of resources and time. It points out that having two separate UN nego-tiating silos on climate change and energy for sustainable development on two intrinsically linked development challenges poses a challenge for any future shared and universal post-2015 development agenda envisaged as an outcome of the 2012 Rio+20 Summit, The chapter briefly touches on a few globally rel-evant, multi-stakeholder alliances/coalitions focused on linking climate change and energy and energy access-related issues. This concluding chapter also argues that while there is absolutely no substitute for the comprehensive emis-sions reductions of CO_2 that are required to limit longer-term climate change impacts, responding to SLCPs (also referred to as short-lived climate forcers such as black carbon [BC] and tropospheric ozone) can also no longer be viewed as controversial and non-actionable because of the derived benefits for human health and poverty reduction. The chapter therefore concludes by iden-tifying regionally appropriate and nationally driven frameworks for action that simultaneously allow for reducing BC emissions and improving access to sustainable energy services for the poor as an area worth exploring, because these mechanisms could directly address the nexus between worsening health impacts of indoor and outdoor air pollution felt by the energy-poor and climate change.

Finally, the book concludes by pointing out that the "siloization" of nego-tiations wastes limited global resources and inhibits integrated action on what are intrinsically linked development challenges. After decades of intergovernmental negotiations on climate change and sustainable development, with their marked proclivity towards parsing negotiating language and texts and convening all-night meetings in diverse cities, key globally agreed outcomes from these two non-intersecting negotiations silos have very little to show in terms of concrete energy-related mitigation and energy access for the poor targets and goals. The crucial role of energy in mitigating anthropogenic climate change and in poverty reduction, as well as the shared 2015 deadline for securing UN-led global agreement on these dual challenges, make ignoring the linkages between sustainable energy access for the poor and climate change increasingly unjustifiable within the broader UN context.

1.3   Escalating risks and increasing costs of the climate change: Scaling up and linkages matter

By all accounts, the issue of global climate change has been heralded as one of the pressing global challenges in diverse fora, by a range of actors and agencies in different countries, and the lack of scientific certainty cannot be referenced as the reason for lack of action on mitigation measures, not just because of the UNFCCC's specific invocation of the precautionary principle but also because of the categorical warnings issued by various UN agencies, and the past and current findings of the IPCC, including its most recent Fifth Assessment Report (IPCC, 2013, 2014). Global climate change happens also to be the only global environmental issue that has been the subject of periodic examination by panels of experts from around the world convened under the aegis of the IPCC. The IPCC has, since 1988, produced a series of five periodic assessments updating knowledge on the scientific, technical, and socio-economic aspects of climate change and has previously concluded that climate change remains one of the greatest challenges facing human society.

Global climate change negotiations are arguably amongst the most complex intergovernmental negotiations, because the issue has, and is anticipated to have, short-, medium-, and long-term impacts on and causal linkages with a wide range of other development concerns, including energy, poverty, health, food security and agriculture, water resources, destruction of ecosystems and infrastructures, and, more tragically, human displacement and loss. Despite all the scientific and policy attention accorded to it, the challenge of finding a global resolution to the problem of human-induced global climate change appears more daunting and fraught with tension today than it was when the UNFCCC was first signed at the historic 1992 Earth Summit in Rio de Janeiro. But, unlike the 1987 Montreal Protocol on Substances that Deplete the Ozone Layer, negotiated under the aegis of UNEP, which had a clearly defined target, namely zero (elimination) production and consumption of ozone-depleting substances (with stated exceptions), the 1992 UNFCC neither referenced nor included any targets for emissions reductions, and it did not immediately obligate or commit countries to any legally binding emissions targets. Crafting and securing a global agreement with buy-in from all the major GHG-emitting countries is further complicated in the case of climate change, because current or temporal vulnerabilities to the adverse effects of climate change are not directly related to historical or current/future aggregate responsibilities for causing the problem, which has made the adoption of lasting or comprehensive targets for GHG emissions reductions hard to achieve.

Annual cycles of intergovernmental negotiations culminating in COPs have occurred ever since the adoption of the historic 1992 UNFCCC, but despite arduous and lengthy negotiating exercises convened in all corners of the globe, a global framework for comprehensive mitigation of GHGs continues to elude the global community, and global climate change remains both an urgent and an intractable global problem. To date, 20 annual COPs have occurred in different parts of the world, and the only agreed mitigation-related outcome to date has been the much-discussed 1997 Kyoto Protocol (KP), with a much more watered-down and less inclusive second phase of the KP agreed to end of 2012. The US and Australia did not sign on to the first period of the KP, and the second phase of the KP has seen a larger group of industrialized developed countries, including Japan, the host country for the first period of KP, indicating their reluctance to join. Not surprisingly, there was no comprehensive global climate change treaty heralded at the 2012 Rio+20 Summit, which met to review 20 years of progress towards sustainable development. At Rio+20, countries could only "reaffirm that "climate change is one of the greatest challenges of our time" and "express profound alarm that emissions of greenhouse gases continue to rise globally" (UNGA, 2012, p. 37).

As evidenced by the summaries of the annual cycles of climate change meetings contained in the Earth Negotiations Bulletin, intergovernmental climate change negotiations have become lengthy, word-parsing negotiating exercises, with multiple and often parallel groups and sub-groups negotiating a diverse range of issues. The structure and format of annual 2-week-long global climate change negotiations focused on securing consensus-based agreements that often cater to the lowest common denominator approach on a vast array of issues has not proved to be conducive to securing a time- and target-bound multilateral response. As the subject of numerous scientific assessments, policy predictions, and diverse political opinions, global climate change has also been the focus of over 20 years of intergovernmental negotiations with no comprehensive global agreement in sight. After more than two decades of around-the-globe UN-led negotiations on global climate change, it is time to ask whether there is a comprehensive energy sector-driven mitigation framework in sight, or whether the annual cycles of climate change negotiations will continue to convene in different cities of the world with diminishing sets of expectations for agreed energy-related mitigation outcomes.

Given the long record of negotiations and globally relevant analyses, this chapter takes as a given that natural and anthropogenic (human influenced) processes and substances (including gases) are the prime drivers of global climate change; it is simply no longer possible to deny human influences on the global climate system; and it is increasingly untenable that the global negotiations fail to provide the means to address and limit such climate change. Global climate change is perhaps the most emblematic of trans-boundary global problems, in that these problems cannot be confined or contained with the boundaries of one state or a set of adjoining states; and, due to their spill-over effects on multiple states, are viewed as requiring multilateral or global responses to ameliorate or address the problem (Cherian, 2012b, p. 39).

Climate change has been demonstrated to be linked to loss of valuable biodiversity and the spread of drought and desertification, which together with climate change comprise the three global environmental conventions called the Rio Conventions that emerged from the historic 1992 Rio Earth Summit. According to an alarming team study published in *Nature* and based on field data on species distribution and regional climate in six biodiversity-rich regions around the world, from Australia to South Africa, the predicted range of climate change by 2050 places 15–35% of the 1,103 species at risk of extinction (Thomas *et al.*, 2004). But the linkages between the loss of endemic and valuable biodiversity resources and the spread of drought and desertification are particularly alarming for small island states and the least developed countries in Africa which are on the frontline of these linked threats.

Devising a global agreement with buy-in from all the major GHG-emitting countries is difficult in the case of climate change because current or temporal vulnerabilities to the adverse effects of climate change are not directly related to historical or current/future aggregate responsibilities for causing the problem. So, for example, what is difficult in terms of allocating responsibility and addressing vulnerability is that the states, such as the least developed countries (LDCs) and small island developing states (SIDS), that have contributed the least in terms of GHG emissions on a per-capita or aggregate basis can also do the least in terms of their responsive capacities to ameliorate the problem and are seen as the most vulnerable frontline states in the struggle to adapt to the adverse effects of climate change (Cherian, 2012a).

Getting a handle on a multidimensional and multi-temporal problem like climate change is very difficult, because there are so many different aspects to consider and address, and to prioritize these various aspects and then devise the requisite responses to them are huge tasks. What the two decades of negotiations evidence is that securing a global consensus-based agreement that will mitigate GHGs (in particular CO_2 emissions associated with fossil fuel use) in a comprehensive and timely manner has proven arduous, and impossible to achieve so far, precisely because individual countries have widely varying responsibilities for this problem, and equally diverse vulnerabilities and responsive capacities toward it. But as this chapter argues, the role of energy in driving human socio-economic development in all forms, and its contribution to the problem of climate change poses a particularly thorny challenge for developing countries with large populations of poor people for whom the policy and programmatic nexus between climate change, energy and poverty reduction is absolutely crucial.

By now, intergovernmental negotiations on global climate change have occurred in diverse fora across the globe for over 20 years at considerable costs, both financial and emissions-related, in terms of transporting and hosting scores of climate change negotiators at different conference and meeting venues. Given the complexity of the global climate change challenge, there have been multiple ways to analyze and approach the issue of intergovernmental climate change negotiations and these are discussed in the following chapter. Climate change negotiations have broadly been focused around two distinct but related issues, but each of these issues is associated with range of serious and complex technology, financing and capacity-related concerns and constraints:

- Mitigation or the reduction of GHG emissions that are seen as principally responsible for the rise in global surface temperatures.
- Adaptation or the human and/or ecosystem-related responses to a range of adverse climatic impacts such as sea-level rise, increase in the frequency and intensity of extreme weather-related events, effects on fragile marine ecosystems, and coastal zone inundation, which accompany a rise in global surface temperatures.

According to the most recent IPCC SPM (Working Group III), which focuses on providing an overall assessment of the status of climate change mitigation, "mitigation" is defined "as a human intervention to reduce the sources or enhance the sinks of greenhouse gases" and "mitigation, together with adaptation to climate change contributes" to the achievement of Article 2 – which is the principal objective of the UNFCCC (IPCC, 2014, p. 3).

Responding to the "absence of tangible outcomes from intergovernmental efforts" to reduce GHG emissions, the 2014 Climate Action in Megacities Report – a quantitative survey assessment released at the February 2014 C-40 Summit in South Africa – notes that 94% of the cities participating in the survey indicate that climate change poses significant risks to their cities: and the report provides evidence of efforts to reduce GHG emission and actions to increase resilience climate risks taken by cities across the world (C-40/Arup, 2014). A more explicit stance was highlighted in the recent WEF report, *Global Risks 2014*, released at the annual gathering of world business and global leaders in Davos, which points out that climate change features among the five most likely and most impactful risks facing the world, the report states that "the risk of climate change by far displays the strongest linkages" to other risks and "can be seen to be both a key economic risk in itself and a multiplier of other risks, such as extreme weather events and water and food crises" (WEF, 2014a).

Adverse climatic impacts are expected to be particularly disruptive, with serious negative socio-economic impacts for developing countries (UNDP, 2007). The stress imposed by climatic impacts on the water–energy–food nexus, which has resulted in drought and water shortages, leading to food shortages and famine, was recently highlighted by the 2014 WEF report on climate change adaptation. In fact, the report points out that "despite the evidence of these linkages" and "the need for an integrated approach to address the impact of climate change", "silo thinking prevails worldwide" and, for instance, solutions to address energy don't take into account water, food or environmental impacts (WEF, 2014b, p. 24). The fact that the world's most vulnerable and poorest regions, which have often contributed the least in terms of per-capita GHG emissions, and also have the least capacities (economic, institutional, scientific, and technical) to cope and adapt to climate change, has proved to be a difficult one to address in the world according to climate change negotiations. But it is one that is critical to the ideas developed in this book, which argues that it is time to have more focused and comprehensive linkages between increasing access to sustainable energy for the poor, and the policy and programmatic synergies between energy access, poverty reduction, and global climate change objectives.

The problem is further complicated by the fact more than half of the world's population now lives in cities. The UN estimated that in 2008, for the first time in history, half of the world's population, or 3.3 billion people, lived in urban areas. One-third of the worldwide urban population, or one billion people, lived in slums and it is estimated that by 2030 at least 61% of the global population will live in cities – with over two billion people living in slums (UNDP, 2010). According to the UN Habitat, most of the increase in urban populations will be in middle- and lower-income countries and the UN estimates that poor populations living in low-lying coastal areas will face the brunt of climatic impacts of rising sea levels, tsunamis, and hurricanes. At the C-40 Cities and Climate Change Summit held in South Africa, the main focus was on cities that are on the frontline of facing the adverse impacts of climate change and in terms of developing solutions. The role of hydrological disasters was made explicit in the *Annual Disaster Statistical Review*, which found that the number of victims of natural disasters increased in 2011 to 244.7 million when compared with an annual average number of 232 million victims from the previous decade – 2001 to 2010. This "increase" was "explained by the larger impact from hydrological disasters", which "caused 139.8 million victims in 2011 – or 57.1% of total disaster victims in 2011 – compared to an annual average of 106.7 million hydrological disaster victims from 2001 to 2010" (Guha-Sapir *et al.*, 2012, p. 1).

Extreme drought and flooding and damage from superstorms could increasingly ravage the economies of poorer countries, locking them more deeply into cycles of poverty. The World Bank estimates the cost of adapting to the adverse impacts of climatic change for developing countries at $70–100 billion per year through to 2050 (World Bank, 2010). It is important to note at the outset that adaptation is related to strengthening developing countries' abilities to respond and build resilience to the adverse impacts of climatic change. But this chapter and this book will not focus on outlining the central role of adaptation efforts or identify adaptation measures and strategies, but will instead focus primarily on the overall lack of specific energy and poverty reduction-related policy references in the two decades of global climate change negotiations. However, it should be clear that linking adaptation efforts with broader national sustainable development objectives, including energy access and poverty reduction efforts, is both urgent and cost-effective, particularly for developing countries where resources (financial and capacity) are limited, and accordingly the relationship between adaptation, energy access and poverty reduction is an important avenue for further consideration.

As reported on by a variety of globally relevant institutions, and endorsed by the global consensus emerging from the climate change negotiations, two time-sensitive factors are worth highlighting in terms of the UN-led process for climate mitigation:

- A scaling up of ambition in terms of climate change mitigation has to take place.
- The window for implementing scaled up action is closing rapidly.

In terms of addressing anthropogenic climate change at its root, the most pressing concern remains the gap between GHG reduction or mitigation pledges and concrete mitigation action and pathways, which makes the possibility of holding the increase in global average temperature to a 2°C limit harder to achieve. UNEP's fourth annual *Emissions Gap Report* warns that "there is a significant gap between the total mitigation pledges of countries in terms of their global annual emissions of greenhouse gases by 2020 and aggregate emissions pathways consistent with keeping the increase in global average temperature below 2°C", but also includes a more cautionary warning that it might already be too late, by stating that "even if the global temperature rise can be held within the 2°C limit, sea levels are expected to continue to rise beyond 2100 since warmth is distributed in the ocean at great depth" (UNEP, 2013a, p. 4). The report also highlighted the urgency to act: delayed action will lead to the need for more costly efforts in the future, or may even make the goal unattainable. Existing commitments from countries play an essential role, but are not sufficient to close the gap. Even under the most stringent models with conditional targets and strict accounting rules, an 8 $GtCO_2e$ (gigatonnes of CO_2 equivalent) gap will remain in 2020 (UNEP, 2013a).

Adding a greater level of pessimism regarding any future closing of the emissions gap, Hulbert (2012a) points out that coal now "accounts for a staggering 30.3% of the global energy mix", which represents the "the largest figure since 1969". According to him, "coal is literally going to be the 'fuel of the future' in Asia", but a series of other emerging market countries will join the "coal growth roster in Asia, Africa and even Latin America" and while this may portend "great news for major exporters (and certain trading houses), [but] from an emissions perspective, it means any chance of keeping global warming below

2°C is close to zero". For him, nothing short of "a miracle in carbon capture & storage (CCS) technology" will be needed, and there will be a shift away from climate change mitigation to climate adaptation by 2020 (Hulbert, 2012a).

The IEA estimates that for each year that passes, the window for action on emissions reductions over a given period becomes narrower. It calculates that each year of delay before moving onto the emissions path consistent with a 2°C temperature increase would add approximately $500 billion to the global incremental investment cost of $10.4 trillion for the period 2010–2030, and that a delay of just a few years would probably render that goal completely out of reach (IEA, 2009, p. 52).

Expressly intended to shock the global community into working with more urgency, a 2012 World Bank report entitled *Turn Down the Heat* states that the emission pledges made at the COPs in Copenhagen (2009) and Cancun (2010), "even if fully met, place the world on a trajectory for a global mean warming of well over 3°C", and that even if these pledges are fully implemented there is still about a 20% chance of exceeding 40°C in 2100 (World Bank, 2012, p. 23). The report is categorical in stating that no nation would be immune to the impacts of climate change in a 4°C world, but points out that in such a world, poverty alleviation efforts would be seriously undermined by the negative impacts of climate change and that "the burden of climate change" will "very likely be borne differentially by those in regions highly vulnerable to climate change" (World Bank, 2012, p. 64). But this idea that the burden of climate change is more negatively borne by the poor has also to take stock of the fact that the poor are already burdened in terms of having constrained access to ineffective and ill-health-engendering energy sources and systems. And it is this double burden of vulnerabilities to climatic impacts and lack of access to sustainable energy that is faced primarily by poor communities in developing countries that is the main frame of reference for the discussions contained below.

But even assuming the fairly optimistic global outcome of holding the temperature increase to 2°C, authors like Guzman (2013) provide growing evidence of negative climatic impacts on human development in terms of severe water and food shortages, famine, and the increased spread of vector-borne diseases. Guzman focuses on the real world and human costs that climate change has had and will continue to have on development, showing that the rise in sea level will inundate small and low-lying nations like the Maldives and densely populated and food-producing coastal areas of Bangladesh, and also showing how long periods of drought in the Sahel have already produced mass violence in Darfur.

Highlighting their grave concern about the "significant gap" between the aggregate mitigation pledges by Parties in terms of global annual emissions of GHGs by 2020 and aggregate emission pathways consistent with having a "likely" chance of holding the increase in global average temperature below

2°C or 1.5°C above pre-industrial levels, negotiators meeting at the 17th annual COP in Durban in 2011 agreed to launch a process applicable to all Parties, and establish an Ad Hoc Working Group on the Durban Platform for Enhanced Action, which would "complete its work as early as possible but no later than 2015" in order to adopt a "protocol, legal instrument or legal out-come at the twenty-first session of the Conference of the Parties and for it to come into effect and be implemented from 2020" (UNFCCC, 2011).

At the 18th annual COP in Doha (December 2012), countries agreed to an extension of the historic 1997 Kyoto Protocol (discussed further in Chapter 2) through 2020, but with a far more limited number of countries signing up to this extension than had previously done so. In the absence of any such agreement, the KP, which was the only measurable outcome of 20 years of climate change negotiations, would have expired days after the Doha round of climate negotiations. Faced with this conundrum, governments agreed to an amendment to KP that would allow for a second commitment period and that also referenced the ongoing role of the Durban Platform with "a view towards ensuring the highest possible mitigation efforts by all Parties" (UNFCCC, 2012).

This reference to "highest possible mitigation efforts by all Parties" holds the key to any future resolution of the global climate change problem, but it also offers a litmus test of sorts for the future of global climate change nego-tiations. After 20 years of negotiations convened all around the globe, it is time to ask if the quixotic quest for the elusive global climate deal will ever end. The question is whether the intergovernmental process will continue to meander along its annual conference cycle with increasingly smaller levels of ambition in terms of mitigation efforts, or whether it will embark on a post-2012 phase focused on scaled-up mitigation efforts. Climate change negotiations were previously held within the context of two distinct working groups, namely the "Ad-hoc Working Group on the Kyoto Protocol" and the "Ad Hoc Working Group on Long-term Cooperative Action", with the later addition of the "Ad Hoc Working Group on Durban Platform for Enhanced Action", which made for very lengthy and complex negotiating sessions. But the fact that intergovernmental negotiations will now coalesce around the Ad Hoc Working Group on the "Durban Platform for Enhanced Action" rather than three separate and parallel negotiating tracks provides a unique historic opportunity to focus on mitigation efforts that can be meaningful to a wide range of countries.

In the past, some, like Chichilnisky and Sheeran (2009), have argued that the two cruel ironies of the climate crisis are: (i) developing countries that are "least responsible for producing" GHGs are now at "greater risk of death and human suffering because of climate change"; and (ii) "for the first time in history, the welfare of rich nations will depend directly on decisions made in poor nations in Africa, Asia and Latin America", because these "nations have the capacity to inflict trillions of dollars of losses in rich countries just by industrializing and using their own fossil fuels as rich countries did and continue to do" (p. 19). But this dichotomization of rich versus poor countries does not reflect current reality, and roles played by the largest aggregate-emitting countries which happen to also be actively pursuing alternative energy pathways to their industrialization.

In fact, the Lima COP-20 outcome related to the submission of INDCs by all Parties can be seen as a global recognition that the scaling up of mitigation efforts will require broader scope of participation from all countries. It is exactly this idea of a more inclusive scope of participation and engagement with global climate change that necessitates that a broader set of goals regarding energy for sustainable development, in particular increasing access to energy for the poor, are more clearly linked to climate change in the UN-led push for more ambitious post-2015 sustainable development agenda.

1.4 Towards a better understanding of the nexus between global climate change and energy access for the poor

Energy is absolutely central to meeting socio-economic development needs at all levels, ranging from the household to the national, and demand for such energy services is increasing rapidly, particularly in large developing countries as they seek economic growth and industrial development. Conversely, the lack of access to cost-effective and reliable energy sources, services, systems, and technologies has well been documented by the UN and other relevant institutions as impacting negatively on socio-economic growth and worsening development challenges, including poverty and health concerns as well as gender inequality at all levels within and across countries. At the global level, the energy sector in all its various manifestations – supply, production, trans-formation, delivery, and use – is a key contributing factor to human-induced climate change. Addressing the linkages between global climate change and energy at the national level is complex and requires multidimensional and multisectoral responses. But the problem of dealing with global climate change and energy gets much more complicated for developing countries that are also dealing with the challenge of poverty reduction and increasing energy access services and systems for the poor.

More than 20 years ago, Flavin and Lenssen (1994) argued that the challenge of climate change would be compounded by the burgeoning energy needs of developing countries whose projected population they estimated at more than six billion in 2020. They pointed out that developing countries have more than three-quarters of the world's population, but their per-capita energy use averages less than one-eighth that in industrial countries; that developing countries' energy use was projected to expand six-fold by 2050; and, that even if per-capita energy use in developing countries were to rise to even one-quarter of the current industrial level by 2025, total energy use would increase by 60% (Flavin and Lenssen, 1994, p. 25). Now with a world population estimated at seven billion, it is unequivocally clear that the global demand for energy has been increasing, led by fossil fuels, which account for over 80% of global energy consumed, a share that has been increasing gradually since the mid-1990s. The energy needs of two of the world's most populous countries will eclipse those of all other developing nations, and together China and India's rise as the world's energy juggernauts has been termed as the "Chindia" energy challenge (Hulbert, 2012b).

Science and data have now confirmed that GHG emissions emanating from energy services and systems are significant contributors to historic increases in atmospheric GHG concentrations, and that the consumption of fossil fuels accounts for the majority of the global anthropogenic GHG emissions (IEA, 2011). As the recent World Energy Outlook Special Report, *Redrawing the Energy Climate Map*, notes, energy lies at the center of climate change, as the energy sector accounts for around two-thirds of GHG emissions and as more than 80% of global energy consumption is based on fossil fuels. Furthermore, governments are not on track to meet their globally agreed goal of limiting long-term rise in the average global temperature to 2°C, whilst global GHG emissions are increasing rapidly, with CO_2 levels in the atmosphere in May 2013 exceeding 400 parts per million for the first time in several hundred millennia (IEA, 2013, p. 9) (see Box 1.2). The IEA report points out that delaying stronger climate action, even to the end of the current decade – 2020 – would be costly because any delay past 2017 would basically lock in the entire carbon emissions budget to 2035 and thereby strand existing energy assets, which would end up costing more in the end to rectify and undo. Their analysis of the entire energy system shows that delaying action on climate change is a false economy, and investments of around $1.5 trillion are avoided in the period to 2020, but an additional $5 trillion of investments are required between 2020 and 2035 to pay, amongst other factors, for the retiring, idling, or retrofitting of carbon-intensive energy assets and infrastructure (IEA, 2013).

At the start of the new millennium, in the context of the UNGA, world leaders agreed to the Millennium Declaration which, for the first time, put the objective of reducing extreme poverty and addressing its many dimensions – hunger, maternal and infant mortality, disease and ill health, lack of infrastructure and shelter, gender inequality, and environmental sustainability – at the front and center of the global development agenda. The adoption of the Declaration resulted in a set of eight time-bound and measurable goals – the MDGs – for combating poverty, hunger, illiteracy, gender inequality, disease, and environmental degradation. The adoption of the MDGs in 2000 represented a critical shift in global development efforts because for the first

> **Box 1.2   The 4-for-2°C policy proposal of the International Energy Agency (IEA, 2013)**
>
> The IEA lists a set of four policies – the 4-for-2°C – that can help keep the door open to the 2°C target through to 2020 at no net economic cost, because they are based on existing, already adopted and proven technologies and measures. Taken together, these policies are estimated to reduce GHG emissions by 3.1 $GtCO_2e$ in 2020 – a whopping 80% of the emission reductions required under a 2°C trajectory, which the IEA report notes would "buy precious time" while the climate negotiations move to Paris in the anticipation of an international agreement. These four policies are worth highlighting as a set of policies on which the UN and the global community could immediately and urgently implement coordinated action:
>
> - Adopting specific energy efficiency measures (49% of the emissions savings).
> - Limiting the construction and use of the least-efficient coal-fired power plants (21%).
> - Minimizing methane (CH4) emissions from upstream oil and gas production (18%).
> - Accelerating the (partial) phase-out of subsidies to fossil fuel consumption (12%).
>
> Source: IEA, (2013); Executive Summary, p. 9.

time, there was a clear focus on putting the poorest and most vulnerable at the heart of the UN's development agenda. But, as Chapter 3 points out, there was no MDG related to increasing energy access for the poor, although the role of energy as a prerequisite for human development and its impacts on human welfare, including access to water, food, health care, education, gender equality, and productive livelihoods, was well known and highlighted. Additionally, there was no explicit linkage or reference linking energy and climate change with poverty reduction in the MDGs even though the historic Kyoto Protocol had been agreed to a mere 3 years prior to this, and climate change had been previously identified by the UN as a key development challenge with impacts on the poor.

In its most recent fourth assessment report of the gap between ambitions/pledges to limit GHGs and concrete emission pathways by which to get to the internationally agreed goal of meeting the 20°C target that emanated from the 2009 Copenhagen COP, UNEP states that relative contributions to global emissions from developing and developed countries changed little from

1990 to 1999, but the balance changed significantly from 2000 to 2010, when the developed country share decreased from 51.8% to 40.9%, while developing country emissions increased from 48.2% to 59.1%. Accordingly, today developing and developed countries are responsible for roughly equal shares of cumulative GHG emissions for the period 1850–2010 (UNEP, 2013b, p. 4). What is relevant to note here, is that while the UNEP report acknowledges at the outset that GHG estimates are uncertain due to differences in definition and in national emissions accounting, energy-related CO_2 emissions have the lowest uncertainty, as compared with land use and land use change-related GHG emissions, which have the highest.

Energy fuels both anthropogenic climate change and human development in all countries across the globe, but lack of access to adequate, sustainable energy services and a heavy reliance on inefficient energy sources contributes to SLCPs that impact negatively on the daily lives and health of millions of poor people and worsens national and regional climate impacts. The heavy dependence on inefficient sources of energy (solid fuels and traditional biomass) and devices (e.g., cook stoves) for heating and cooking purposes in poor households in developing countries has been documented to result in incomplete combustion and the release of SLCPs, including BC, which has serious health and climate change impacts (UNEP, 2013b, p. 4). In particular, emissions that result from the inefficient burning of traditional forms of biomass on which poor households rely are released as a mix of health-damaging pollutants, such as BC, that have short atmospheric life spans but make significant impacts to human health and climate change at regional and global levels (UNEP/World Meteorological Organization [WMO], 2011). The key issue is whether and how any future UN-led energy-related mitigation framework will respond to the needs of millions who lack access to energy to service their basic needs.

Time is literally running out for securing a global consensus-based comprehensive climate change mitigation agreement that can ensure a 2°C world for the future, with the Paris COP-21 being billed as the global venue for a much anticipated ambitious and inclusive climate change agreement. The question is whether the 2015 agreed climate change options will include concrete references related to increasing access to sustainable energy services for the poor and the mitigation of SLCP emissions accruing from inefficient energy sources and systems.

There is no doubt that mitigating energy-related GHG emissions lies at heart of any future and effective resolution of global climate change, but continuing to ignore the linkages between increasing access to sustainable energy and its attendant benefits on poverty reduction, human health and SLCP mitigation contravenes the broader UN quest for an integrated and universal post-2015 development agenda.

Even though the issue of SLCPs has not been the subject of any formal intergovernmental climate change negotiated outcomes, it is important to point out that influential assessments by UNEP and WMO (2011) have highlighted the role of SLCPs. More recently, Working Group 1 of AR5 has found larger contributions to climate change from methane and carbonaceous aerosols than the previous Fourth IPCC assessment. AR5 clearly distinguishes between what is referred to as "well-mixed greenhouse gases" (CO_2, CH_4, nitrous oxide, and halocarbons) and "near term climate forcers" such as BC and methane. With methane being included in both categories, the IPCC's Working Group I report defines:

> … 'near-term climate forcers' (NTCFs) as those compounds whose impact on climate occurs primarily within the first decade after their emission. This set of compounds is composed primarily of those with short lifetimes in the atmosphere compared to WMGHGs, and has been sometimes referred to as short-lived climate forcers or short-lived climate pollutants. (IPCC, 2013, p. 668)

Figure 1.3, which is excerpted from the IPCC AR5 SPM report of Working Group I, provides a schematic of the radiative forcing estimates in 2011 for the main drivers of climate change.

The SPM report of the IPCC's Fifth Assessment Report points out that "the largest contribution to total radiative forcing is caused by the increase in the atmospheric concentration of CO_2 since 1750" and distinguishes between the climate impacts of long-lived GHGs such as CO_2 and near-term climate forcers and does not endorse or encourage the use of CO_2 equivalence or the use of a single metric or time horizon as somehow more preferable for characterizing the impacts of emissions (IPCC, 2013, p. 13). Emissions reductions of near- or short-term climate forcers cannot therefore substitute for the agreed-upon emissions reductions of GHGs, which are required to limit anthropogenic climate change, but simply ignoring the impacts of indoor air pollution and the role of SLCPs that result from the heavy reliance and use of inefficient energy sources, including traditional biomass, has devastating development consequences for millions of people. In addressing the nexus between energy poverty and climate change mitigation, it is therefore useful to consider the impacts of all drivers of climate change, including the role of near-term climate forcers such as BC, as highlighted by the figure contained in AR5 of WG1. The linkages between the energy sector and climate change are clearly underscored by this SPM report which states that:

> A single year's worth of current global emissions from the energy and industrial sectors have the largest contributions to global mean warming over the next approximately 50 to 100 years. Household fossil fuel and biofuel, biomass burning and on-road transportation are also relatively large contributors to warming over these time scales. (IPCC, 2013, p. 663)

The reality is that there has been a considerable amount of research on implications of "energy poverty", but these analyses have not translated

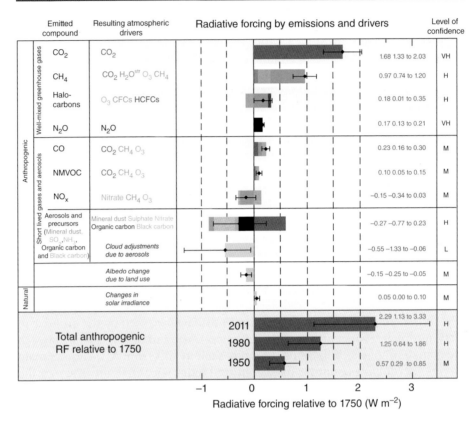

Figure 1.3 Radiative forcing estimates in 2011 relative to 1750 and aggregated uncertainties for the main drivers of climate change. Values are global average radiative forcing (RF14), partitioned according to the emitted compounds or processes that result in a combination of drivers. The best estimates of the net radiative forcing are shown as black diamonds with corresponding uncertainty intervals; the numerical values are provided on the right of the figure, together with the confidence level in the net forcing (VH, very high; H, high; M, medium; L, low; VL, very low). Albedo forcing due to black carbon on snow and ice is included in the black carbon aerosol bar. Small forcings due to contrails (0.05 W m^{-2}, including contrail-induced cirrus), and hydrofluoro-carbons (HFCs), perfluorocarbons (PFCs), and sulfur hexafluoride (SF6) (total 0.03 W m^{-2}) are not shown. Concentration-based RFs for gases can be obtained by summing the like-colored bars. Volcanic forcing is not included as its episodic nature makes it difficult to compare to other forcing mechanisms. Total anthropogenic radiative forcing is provided for three different years relative to 1750 (source: IPCC, 2013, figure SPM.5).

into globally negotiated outcomes within the UN context. For instance, using income distribution and national energy data at the individual level, Chakravarty and Tavoni (2013) argue that an energy poverty alleviation program can provide for basic energy needs of the poor while raising energy consumption by only 7% with additional emissions estimated at 44–183 GtCO$_2$ over the 21st century with a maximum warming contribution of only 0.13°C. Meanwhile, Ürge-Vorsatz and Herrero (2012) compiled a taxonomy of policy interactions that argue that solving both global problems – climate change and poverty reduction – requires the integration of policy goals. The urgent need for policies that simultaneously address climate change and the energy needs of the poor was identified as a central challenge in a paper calling for a comprehensive program on clean energy and increased inter-sectoral research and action (Haines *et al.*, 2007). Linkages between energy and climate change have been considered, but not in terms of linking two intergovernmental processes and providing specific energy sector climate mitigation options for consideration as this current book does. For example, Dodds *et al.* (2009) has examined the "twin" challenges of climate change and the energy supply crunch. But here the focus has been on the challenges of meeting future energy demands and energy security and on how global climate change poses a threat to global security, peace, and development.

Further complicating the ongoing global discussion on allocating responsibilities for addressing anthropogenic GHG emissions is the finding that cumulative GHG emissions from large developing countries (e.g. the BASIC countries) are growing rapidly. And there is an additional and relatively new challenge faced by large developing countries like China and India, which are also amongst the world's large aggregate GHG emitters: worsening air pollution and health impacts resulting primarily from incomplete combustion of inefficient fuel sources and systems. According to Harris (2015), while Beijing's air quality may have garnered more headlines, New Delhi's air "is the world's most toxic in part because of the high concentrations of PM 2.5", especially during the winter months due to the widespread burning of garbage, coal, and diesel fuel, and an estimated 1.5 million people, comprising approximately one-sixth of all national deaths, die annually as a result of both indoor and outdoor pollution.

Heck and Hirschberg (2011) estimate the external costs of air pollution in China to range from approximately 1% to 8% of China's gross domestic product (GDP). They argue that despite the discrepancies that arise due to uncertainties in impact modeling, there is consensus that damages from air pollution in China add up to a substantial share of GDP. The issue of addressing air pollution and climate change in the largest of developing countries – China – is also considered by Kan *et al.* (2012) who note that climate change has negatively impacted on human health in China both directly and indirectly, and argue that a focused consideration of the serious health impacts of air pollution and climate change should inform China's action towards reducing fossil fuel combustion and resultant air pollution, and assist in moving towards sustainable development.

Increases in outdoor air pollution combined with growing ill-health and deaths caused by the heavy reliance on inefficient energy systems and badly combustible solid fuels amongst the poor are a huge threat to public health, as demonstrated by the most recent WHO estimates on air pollution in developing countries. The WHO has long been sounding the alarm on the public health impacts of indoor air pollution due to a lack of access to clean and sustainable energy services at the household level. According to the WHO, in "2000, indoor air pollution was responsible for more than 1.5 million deaths and 2.7% of the global burden of disease", but "in high-mortality developing countries, it accounted for 3.7% of the burden of disease, disproportionately impacting on women and children who spend the most time near the domestic hearth" (WHO, 2007, p. 1). Troublingly, there has been an exponential growth in terms of the health and morbidity impacts of air pollution. In its newest 2014 report, on the burden of disease caused by indoor and outdoor air pollution released in March 2014, the WHO pointed out that "globally, 7 million deaths were attributable to the joint effects of household air pollution (HAP) and ambient air pollution (AAP) in 2012", with "the Western Pacific and South East Asian regions bearing most of the burden with 2.8 and 2.3 million deaths", respectively, with "680,000 deaths which occur in Africa"(WHO 2014a, p. 1). In comparison, though, the costs of indoor air pollution are greater and felt more directly by the poor and here the WHO has found that "globally, 4.3 million deaths were attributable to household air pollution (HAP) in 2012, almost all in low and middle income countries", and where "South East Asian and Western Pacific regions bear most of the burden with 1.69 and 1.62 million deaths respectively" with almost "600,000 deaths which occur in Africa" (WHO, 2014b, p. 1).

The costs of energy-related air pollution for developing countries are becoming increasingly harder to ignore. Measures such as increasing access to sustainable energy services and improving access to innovative sustainable energy technologies could reduce air pollution and save millions of lives. The issue of new forms of partnerships and innovative frameworks for action that link climate change and energy access for the poor will be discussed further in Chapter 5. Box 1.3, based on information from WHO, provides stark new evidence of the seriousness and scale of the problem of HAP in poor households that the WHO terms as the "forgotten three billion".

It is important to note that several additional studies have demonstrated the multiple health and environment benefits that result from improving access to clean and cost-effective energy services for the poor. Based on a modeling framework that integrates indoor and outdoor air pollution legislations and models, Rao *et al.* (2013) argue in favor of a combined or synergistic approach of policies on outdoor air pollution, climate change and access to clean cooking fuels can result in a significant decline in the global burden of disease from both outdoor and household air pollution. Along the same lines,

> ### Box 1.3   Key facts on household air pollution and health
>
> - Around three billion people still cook and heat their homes using solid fuels (i.e. wood, crop wastes, charcoal, coal and dung) in open fires and leaky stoves. Most are poor, and live in low- and middle-income countries.
> - Such inefficient cooking fuels and technologies produce high levels of household air pollution with a range of health-damaging pollutants, including small soot particles, that penetrate deep into the lungs.
> - In poorly ventilated dwellings, indoor smoke can be 100 times higher than acceptable levels for small particles.
> - Exposure is particularly high among women and young children, who spend the most time near the domestic hearth.
> - Over four million people die prematurely from illness attributable to the household air pollution from cooking with solid fuels.
> - More than 50% of premature deaths among children under 5 are due to pneumonia caused by particulate matter (soot) inhaled from household air pollution.
> - 3.8 million premature deaths annually from non-communicable diseases, including stroke, ischemic heart disease, chronic obstructive pulmonary disease and lung cancer, are attributed to exposure to household air pollution.
>
> Source: WHO Media Centre (2014).

van Vliet *et al.* (2012) argue that combining air pollution control policies, energy access, and climate policies can further help to reduce both outdoor and indoor air pollution-related health impacts in Asian energy systems. They point out that investments into energy systems must double by 2030 to achieve climate goals but that strong end-use efficiency and other measures help to lower costs and that the costs of energy access policies are a fraction of the total energy system costs.

In addition to the health-related costs of pollutants that result from inefficient cook stoves and energy sources, there is the additional concern about emissions of BCs that accrue from burning of solid fuels that was discussed briefly in the introductory section of this chapter. The release of particulate matter less than 2.5 micrometers in diameter (PM 2.5), which is a component of BC emitted as a result of incomplete combustion of solid fuels, biomass including by diesel-based sources, cook stoves and open fires, has been quite extensively highlighted by UNEP.

A recent UNEP report outlining key emerging risks to the global environment pointed to the negative health and environmental impacts of BC particles, noting that:

> These particles are emitted as a result of incomplete combustion of fuel and bio-mass, for example by diesel vehicles, cooking and heating stoves and open fires... PM2.5 is a component of black carbon. Black carbon and ground level ozone (or tropospheric ozone are two of the short lived climate pollutants (SLCPs). Apart from making up a substantial part of the air pollutants that impact human health, SLCPs contribute to climate change... Fast and sustainable action to reduce emissions of SLCPs represents a major opportunity to deliver multiple benefits in terms of public health, food and energy security and near term climate protection (UNEP, 2013b, p. 4).

With regard to the critical role of SLCPs, it important to also highlight an influential and comprehensive joint 2011 UNEP/WMO assessment that identified BC and tropospheric ozone (which are not GHGs) along with methane (which is also an important GHG) and some hydrofluorocarbons (HFCs) as the four crucial short-lived climate forcers that are fundamentally different from longer-lived GHG gases such as carbon dioxide, and that these SLCPs (also referred to as short-lived climate forcers) remain in the atmosphere for a much shorter time period. This joint assessment is categorical in pointing out that "a small number of emissions reduction measures targeting black carbon and tropospheric ozone precursors (such as methane and carbon monoxide) could immediately begin to protect climate, public health, water and food security and ecosystems", but also categorical in pointing out that "deep and immediate cuts in carbon dioxide reductions are required to protect long-term climate as this cannot be achieved by addressing short lived climate forcers" (UNEP/WMO, 2011, p. 2). The finding that multiple benefits accrue from a small set of identified measures merits close and urgent consideration and will be discussed further in Chapter 5.

At this juncture, the issue of addressing the nexus between lack of energy access for the poor and climate change mitigation is not a binary question of whether or not actions to reduce BC and tropospheric ozone precursors are somehow more or less important than longer-term actions to mitigate GHGs, because this poses a false dichotomy of pitting longer-term climate mitigation against short-term mitigation of SLCPs. Both long-term mitigation of GHGs and short-term climate mitigation of SLCPs are necessary, but there are two crucial current caveats to keep in mind:

1. There are significant uncertainties associated with the issue of emissions reductions of SLCPs and there are no formally agreed-upon measures relating to SLCPs within the context of current intergovernmental climate change negotiations.
2. Long-term mitigation strategies will need to be very much part and parcel of the 2015 global climate mitigation agreement.

To put it in somewhat simple yet stark terms, what appears to be fundamentally at issue is how to maintain and/or who has control over the global space for economic development fueled by the growing energy sector and its attendant GHG emissions and how to and/or who will pay the costs in terms of lives lost, ill health, and financial resources for mitigating and adapting to global climate change. While not all nations are equally responsible for cumulative and/or historical atmospheric GHG emissions, unfortunately the polluting and ill-health-engendering nature of SLCPs emissions have a more direct impact on poorer populations in the countries and regions where they emanate.

As discussed earlier, many global reports have stressed that the poor in developing countries who have contributed the least in terms of per-capita GHG emissions will suffer the most as a result of the adverse impacts of climate change. But what is often missing from the analyses has been linkages to the double burden faced by the poor in developing countries, in terms of the devastation wrought from adverse climate change impacts as well as the health impacts of SLCPs that result from their heavy dependence on inefficient energy sources. Unfortunately, the global quest for a sustainable energy future is not exactly borne out in terms of the UN-led global negotiations on climate change and sustainable development, which, as Chapter 2 and 3 will evidence, are conducted on distinct and separate tracks with no real programmatic and policy guidance relating to a framework for action on sustainable energy access for the poor.

1.5 Energy access for the poor and climate change cannot exist as separate global silos: A shared post-2015 development agenda necessitates integration not siloization

Energy is fundamental to addressing anthropogenic GHG emissions, and access to sustainable energy is absolutely vital to reducing poverty, which remains both the overarching goal of the global development community and the immediate concern of a majority of developing countries as they discuss the post-2015 sustainable development agenda in the UN. As Chapter 2 demonstrates, there has been a great deal of analysis focused on the global climate change regime and climate change negotiations, but focused attention has not been paid to the idea that currently there are fractured and separate global negotiating tracks of energy for sustainable development and global climate change which have serious implications for the broader global development efforts to reduce poverty and for the engagement of developing countries that are simultaneously faced with addressing the lack of energy services whilst also finding the need to respond to climate change objectives.

Clearly, finding a timely and comprehensive global resolution to the climate change challenge has become harder when countries across the globe are faced with the impacts of the global economic crisis and burdened with growing energy costs. The search for global consensus that would respond to the effects of growing national energy demands, and mitigate against increasing GHG emissions, in the face of socio-economic needs, poverty reduction concerns, and escalating vulnerabilities to adverse climatic impacts is nothing short of a Herculean task. But it is one that is made even more challenging in the absence of developing innovative modalities and mechanisms that put energy sector issues front and center in any future global climate change negotiations. Increasing energy consumption from a variety of conventional and non-conventional fuel sources is crucial for fueling economic development in all countries. Increasing quantities of GHGs, including CO_2, methane and nitrogen oxides, resulting primarily from fossil fuel burning and land use changes, have been and continue to be emitted into the Earth's atmosphere. It has been estimated that GHG emissions from developing countries are likely to exceed those from developed countries within the first half of this century (IEA, 2009, 2011). But there is a substantive policy disconnect in addressing energy sector issues in a concrete manner within the context of the intergovernmental negotiations on climate change, which is puzzling in light of the global consensus that energy sector-related emissions play a crucial role in anthropogenic global climate change.

The global community has to take stock of the annual cycles of climate change negotiations conducted over a period of 20 years, which have been on a long quest to find an effective and comprehensive framework agreement on mitigation, but in which energy for sustainable development and energy access for the poor have not really been referenced. Energy issues are central not just the future resolution of the problem of increasing anthropogenic global climate change, but also to the broader UN search for a more sustainable future and its clearly stated development priority to reduce poverty. The absence of reliable access to clean energy services imposes a large disease burden on poor countries and populations and also contributes to worsening environmental conditions. The central development challenge in terms of the energy–poverty–climate change nexus is to enable the global community to simultaneously prevent dangerous anthropogenic climate change while also addressing the need to increase access to clean energy for the poor. Ignoring policy and programmatic linkages and references between climate change, energy access and poverty reduction does a disservice to the broader UN search for sustainable development, whilst focusing on this nexus provides synergies amongst three existing development-related concerns that are pertinent to the needs of developing countries.

Existing analyses and programmatic work in developing countries conducted by major intergovernmental organizations, including the World

Bank, the UNEP and UNDP, indicate that the relationship among energy access, poverty reduction and climate change is a crucial development linkage. For all countries, but particularly for developing ones dealing with the challenges of reducing poverty whilst also fueling economic growth, it is precisely the policy and programmatic linkages between increasing access to energy services, poverty reduction and climate change that are absolutely critical to achieving the broader goal of sustainable development. Addressing linkages between global climate change and increasing energy access to reduce poverty is no doubt complicated, but they are absolutely central to the broader UN-led global quest towards post-2015 sustainable development and should not be ignored or separated out in terms of negotiated outcomes and/or kept in distinct negotiating tracks or silos.

The prime impetus in calling for an integrated framework for intergovernmental negotiations on climate change and energy for sustainable development, is that such a framework would promote synergies and benefit poor and marginalized communities across the globe, who not only suffer from a lack of access to sustainable energy services, but are also in the frontline in terms of coping with tragic and extreme climate-induced weather events, including superstorms and typhoons. The potential for focusing on linkages between increasing energy access for the poor and global climate change mitigation is also timely and urgent given the ongoing UN-led high level global negotiations that aim to define the post-2015 development agenda. The lack of guidance and inputs within the context of the separate UN-led negotiations on climate change and energy development on innovative modalities and mechanisms that explicitly link energy access and poverty reduction and put them front and center in any future global climate change negotiations can therefore be seen as a missed opportunity – one that limits action on a more inclusive and inter-linked post-2015 sustainable development agenda.

If the Durban Platform's objective of scaling up mitigation efforts is to be realized by the 2015 UN Climate Change COP to be held in Paris, then any future integrated climate change and energy access negotiations should focus on promoting the role of innovative sustainable energy technologies that can address the needs of the poor. But it should also take stock of recent scientific analyses related to emissions reductions actions that focus on near-term strategies, such as those that can be achieved by addressing short-lived climate forcers such as BC particles (component of soot), which is not a GHG but has a strong warming effect and, more significantly, impacts on human health (UNEP/WMO, 2011). As discussed further in Chapter 5, the role of these short-lived climate forcers cannot ever substitute the role of anticipated and agreed-on comprehensive emissions reductions of GHGs, including carbon dioxide, which are required to limit longer-term climate change impacts, nor are they to be viewed as a mechanism to shift the historical responsibility for anthropogenic climate change.

Bold and prompt global action is urgently needed to face up to the dual and intersecting development challenges of increasing energy access for the poor and also addressing global climate change. There are two reasons why closer programmatic and policy linkages between the energy access for the poor and climate change make sense at this juncture within the broader UN context:

1. The UN-led process and structure of the climate change negotiations has become cumbersome, unwieldy, and divisive over the course of the two decades. At the same time, there is a parallel but separate UN-led process which is discussing energy for sustainable development goals and objectives. Having two separate negotiating silos on intersecting development challenges that result in two distinct sets of outcomes with minimal linkages is not an effective use of limited resources and capacities.
2. Mitigating energy-related GHG emissions lies at heart of any future resolution of global climate change, and increasing access to modern and clean energy is essential to achieving poverty reduction. Energy for sustainable development objectives, including increasing access to sustainable energy services for the poor, is a crucial integrative element bringing together sustainable development, poverty reduction, and climate change-related development agendas.

As a follow up to the historic 2012 Rio+20 Conference and its outcome document (UNGA, 2012), on May 30, 2013, the High Level Panel of eminent persons on the Post-2015 Development Agenda tasked by the UN Secretary-General reported on the importance of developing a single agenda by 2015:

> Member states of the General Assembly of the United Nations have also agreed at Rio+20 to develop a set of sustainable development goals that are coherent with and integrated into the development agenda beyond 2015. 2015 also marks the deadline for countries to negotiate a new treaty to limit greenhouse gas emissions. Developing a single, sustainable development agenda is critical. Without ending poverty, we cannot build prosperity; too many people get left behind (UN, 2013, p. 5).

In the 2014 UN Secretary-General's Synthesis Report on the post-2015 development agenda, entitled *The Road to Dignity: Ending Poverty, Transforming All Lives and Protecting The Planet*, 2015 has been billed as a historic year in which the global community has been called to action to undertake an extremely ambitious set of 17 sustainable development goals (SDGs). The listing of these 17 goals was the result of a lengthy process of global negotiations and consultations and agreed on as part of an UN open-ended group convened on securing agreement on a set of comprehensive SDGs that would govern the post-2015 sustainable development agenda. These SDGs, in particular Goal 7 (energy access) and Goal 13 (climate change), are discussed in more detail in Chapter 3 as well as in the last two chapters. Interestingly, this 2014 UN synthesis report notes that:

> Tackling climate change and fostering sustainable development agendas are two mutually reinforcing sides of the same coin. To achieve these ends, all have called for a **transformational** and **universal** post-2015 agenda, buttressed by science and evidence, and built on the principles of human rights and the rule of law, equality and sustainability (emphasis included) (UNGA, 2014, para 49, p. 11).

But, as the remaining chapters demonstrate, the questions facing the UN global community in 2015 and beyond are whether increasing access to sustainable energy services for the poor will be explicitly linked to, and reinforced by, the proposed 2015 global climate change agreement, or whether these two inherently linked development challenges will continue to kept separate and distinct. It is not clear what the final 2015 anticipated agreement on global sustainable development goals will include in terms of concrete energy access and climate change targets and goals, and whether concrete programmatic synergies between energy and climate change will be implemented as part of the post-2015 development agenda, including any anticipated 2015 global climate change agreement achieved at COP-21 in Paris. Preserving the existing negotiated silos on climate change and energy access for the poor cannot be construed as "bold action" to address these dual and intersecting problems.

References

African Development Bank, Asian Development Bank, Department for International Development *et al.* (2003) *Poverty and Climate Change: Reducing the Vulnerability of the Poor through Adaptation.* http://www.oecd.org/env/cc/2502872.pdf[accessed May 11, 2015].

Bazilian, M., Nussbaumer P., Haites E. *et al.* (2010) Understanding the scale of investment for universal energy access. *Geopolitics of Energy* (Special Issue: Energy Poverty and Development), 32(10), 21–42.

Blunden, J. and Arndt, D.S. (eds) (2014) State of the Climate in 2013. *Special Supplement to the Bulletin of the American Meteorological Society*, 95(7), S1–S257.

Brew-Hammond, A. (2010) Energy access in Africa: challenges ahead. *Energy Policy*, 38(5), 2291–2301.

C-40/Arup (2014) *Climate Action in Megacities* (Vol. 2.0). http://issuu.com/c40cities/docs/c40_climate_action_in_megacities[accessed May 11, 2015].

Chakravarty, S. and Tavoni, M. (2013) Energy poverty alleviation and climate change mitigation: Is there a trade off? *Energy Economics*, 40, Suppl. 1. S67–S73.

Cherian, A. (2012a) Grappling with the global climate challenge. In: Harris, F., ed. *Global Environmental Issues.* Oxford: Wiley-Blackwell Publications, pp. 65–86.

Cherian, A. (2012b) Confronting a multitude of multilateral environmental agreements. In: Harris, F., ed. *Global Environmental Issues.* Oxford: Wiley-Blackwell Publications, pp. 39–61.

Chichilnisky, G. and Sheeran K. (2009) *Saving Kyoto*. London: New Holland Publishers.

Davenport, C. (2015) "A climate accord based on global peer pressure". *New York Times,* December 14, 2014.

Dodds, F., Higham, A. and Sherman, R. (eds) (2009) *Climate Change and Energy Insecurity: The Challenge for Peace, Security and Development*. London: Earthscan.

Flavin, C. and Lenssen, N. (1994) *Power Surge: Guide to the Coming Energy Revolution*. New York: W.W. Norton.

Goldemberg, J., Johansson, T.B., Reddy, A.K.N. and Williams, R.H. (1988) *Energy for a Sustainable World*. New Delhi: Wiley Eastern Limited.

Gore, A. (2006) An Inconvenient Truth: The Planetary Emergency of Global Warming and What We Can Do About It. New York: Rodale.

Green Climate Fund (GCF). http://www.gcfund.org/about/mandate.html[accessed April14, 2014].

Guha-Sapir D., Vos F, Below, R. and Ponserre, S. (2012) *Annual Disaster Statistical Review 2011: The Numbers and Trends*. Brussels: CRED.

Guzman, A. (2013) *Overheated: The Human Cost of Climate Change*. New York: Oxford University Press.

Haines, A., Smith, K.R., Anderson, D. *et al*. (2007) Policies for accelerating access to clean energy, improving health, advancing development, and mitigating climate change. *The Lancet*, 370(9594), 1264–1281.

Harris, G. (2015) "Delhi wakes up to an air pollution problem it cannot ignore", *New York Times*, February 14, 2015.

Heck. T. and Hirschberg, S. (2011) China: economic impacts of air pollution in the country. In: Nriagu, J.O., ed. *Encyclopedia of Environmental Health*. Burlington: Elsevier, pp. 625–640.

Hulbert, M. (2012a) India's blackouts and the dark future of emerging markets: more disruptions, more coal. *Forbes*. http://www.forbes.com/sites/matthewhulbert/2012/08/01/indian-blackout-three-inconvenient-truths/ [accessed May 18, 2014].

Hulbert, Matthew (2012b) "'Chindia' energy bloc rising? *Forbes*. http://www.forbes.com/sites/matthewhulbert/2012/06/21/chindia-energy-bloc-rising/ [accessed March 12, 2014].

IEA (2009) *World Energy Outlook: 2009*. Paris: OECD/IEA.

IEA (2011) *World Energy Outlook: 2011*. Paris: OECD/IEA.

IEA (2013) *Redrawing the Energy-Climate Map: Special Report*. Paris: OECD/IEA

IPCC (2012) *Summary for Policymakers: Managing the Risks of Extreme Events and Disasters to Advance Climate Change Adaptation*. A Special Report of Working Groups I and II of the IPCC. Cambridge: Cambridge University Press.

IPCC (2013) *Summary for Policymakers in Climate Change 2013: The Physical Science Basis*: Contribution of Working Group I to Fifth Assessment Report (AR5) of the Intergovernmental Panel on Climate Change. Cambridge: Cambridge University Press.

IPCC (2014) *Summary for Policymakers: Working Group III*. Contribution of Working Group III to Fifth Assessment Report (AR5). http://report.mitigation2014.org/spm/ipcc_wg3_ar5_summary-for-policymakers_approved.pdf [accessed May 11, 2015].

IUCC (1992) *United Nations Framework Convention on Climate Change*. Geneva: UNEP/WMO IUCC.

Johnson, T. (2007) *Major Emitters Go To Washington. Council on Foreign Relations: Analysis Brief.* www.cfr.org/international-organizations-and-alliances/major-emitters-go-washington/p14317 [accessed August 20, 2013].

Kan, H., Chen, R. and Tong, S. (2012) Ambient air pollution, climate change, and population health in China, *Environment International*, 42, 10–19.

Ki-Moon, Ban (2012) "Powering sustainable energy for all" *New York Times,* January 11, 2012.

MEF website, About the MEF. http://www.majoreconomiesforum.org/about.html [accessed May 22, 2014].

Modi, V., S. McDade, D. Lallement, and Saghir, J. (2006) *Energy and the Millennium Development Goals.* New York: Energy Sector Management Assistance Programme, UNDP/UN Millennium Project/World Bank.

Montes, M. (2012) Understanding the Long Term Finance Needs of Developing Countries (South Centre: presentation to the first workshop on long term finance). http://unfccc.int/files/cooperation_support/financial_mechanism/long-term_finance/application/pdf/montes_9_july_2012.pdf [accessed May 11, 2015].

NOAA (2015) 2014 Annual Climate Report. http://www.ncdc.noaa.gov/sotc/global/2014/13 [accessed January 16, 2015].

Oberthur, S. and Ott, H.E. (eds) (1999) *The Kyoto Protocol: International Climate Policy for the 21st Century.* Berlin: Springer.

OECD/IEA (2011) *Energy for All: Financing Energy Access for the Poor.* Special Report on the World Energy Outlook. Paris: OECD/IEA.

Office of the US Press Secretary (2009) Declaration of the Leaders of the Major Economies Forum. http://www.whitehouse.gov/the_press_office/Declaration-of-the-Leaders-the-Major-Economies-Forum-on-Energy-and-Climate/ Accessed on May 22, 2014.

Parnell, J. (2013) Green Climate Fund Set to open for business in 2014. RTCC. http://www.rtcc.org/2013/07/02/green-climate-fund-set-to-open-for-business-in-2014/ [accessed May 11, 2015].

Rao, S., Pachauri, S., Dentener, F., *et al.* (2013) Better air for better health: Forging synergies in policies for energy access, climate change and air pollution. *Global Environmental Change*, 23(5), 1122–1130.

Rehman, I.H., Kar, A., Banerjee, M. *et al.* (2012) Understanding the political economy and key drivers of energy access in addressing national energy access priorities and policies. *Energy Policy*, 47(1), 27–37.

Sokona, Y., Sarr, S. and Wade, S. (2004) Energy Services for the Poor in West Africa: Sub-regional "Energy Access" Study of West Africa. Global Network on Energy for Sustainable Development. http://www.gnesd.org/~/media/Sites/GNESD/Publication%20pdfs/Energy%20Access%20theme/Technical_report_ENDA_ver_16_April_2004.ashx?la=da [accessed May 11, 2015].

Sokona, Y., Mulugetta, Y. and Gujba, H. (2012) Widening energy access in Africa: Towards energy transition. *Energy Policy*, 47(1), 3–10.

Sovacool, Benjamin (2012) The political economy of energy poverty: A review of key challenges, *Energy for Sustainable Development*, Vol. 16: 3. p. 272-282

Stern Review (2006) *Stern Review on the Economics of Climate Change.* London: HM Treasury.

Thomas, C., Cameron, A. Green, R.E. *et al.* (2004) Extinction risk from climate change. *Nature*, 427, 145–148.

UN (2013) *The New Global Partnership: Eradicate Poverty and Transform Economies through Sustainable Development.* New York: UN.

UN (2014) *Climate Summit 2014: Action Areas/Summit Announcements.* http://www.un.org/climatechange/summit/action-areas/ [accessed September 27, 2014].

UN Climate Summit/Executive Office of the UN Secretary-General (2014) Build Consensus and Implement Actions For a Cooperative and Win-Win Global Climate Governance System Sept 23, 2014, Statement by H.E Zhang Gaoli. http://statements.unmeetings.org/media2/4628014/china_english.pdf [accessed on October 1, 2014].

UNDESA (2007) *The Effects of Power Sector Reform on Energy Services for the Poor.* New York: UN.

UNDP (2007) *Human Development Report 2007/2008: Fighting Climate Change.* New York: UNDP.

UNDP (2011) *Towards an 'Energy Plus' Approach for the Poor.* New York:UNDP.

UNDP/UNDESA/WEC (2000) *World Energy Assessment: Energy and the Challenge of Sustainability.* New York: UNDP.

UNDP/WHO (2009) *The Energy Access Situation in Developing Countries.* New York: UNDP/Environment and Energy Group.

UNEP (2012) *Emissions Gap Report 2012* (Nairobi: UNEP, 2012).

UNEP (2013a) *Emissions Gap Report 2013.* Nairobi: UNEP

UNEP (2013b) *Year Book: Emerging Issues in Our Global Environment.* Nairobi: UNEP.

UNEP/WMO (2011) *Integrated Assessment of Black Carbon and Tropospheric Ozone: Summary for Decision Makers.* Nairobi: UNEP.

UNFCCC (2008) *Climate Change: Impacts, Vulnerabilities and Adaptation in Developing Countries.* Bonn: UNFCCC Secretariat.

UNFCCC (2011) Establishment of an Ad Hoc Working Group on the Durban Platform for Enhanced Action, FCCC/CP/2011/L.10. December 10, 2011.

UNFCCC (2012) Outcome of the work of the Ad Hoc Working Group on Further Commitments for Annex I Parties under the Kyoto Protocol. FCCC/KP/CMP/2012/L.9.

UNFCCC (2015) Report of the Conference of the Parties on its twentieth session held in Lima from 1-14 December 2014. Decisions adopted by the Conference of the Parties: Addendum 1. FCCC/CP/2014/Add.1. http://unfccc.int/resource/docs/2014/cop20/eng/10a01.pdf [accessed February 17, 2015].

UNFCCC/ADP (2015) Ad Hoc Working Group on the Durban Platform for Enhanced Action Second Session Part 8: Negotiating text. FCCC/ADP/2015/1. http://unfccc.int/resource/docs/2015/adp2/eng/01.pdf [accessed April 22, 2015].

UNFCCC website (2015) Green Climate Fund. http://unfccc.int/cooperation_and_support/financial_mechanism/green_climate_fund/items/5869.php [accessed on January 12, 2015].

UNFCCC/INDC website (2015) INDCs as communicated by the Parties. http://www4.unfccc.int/submissions/indc/Submission%20Pages/submissions.aspx [accessed April 22, 2015].

UNGA (1988) *Protection of Global Climate Change for Present and Future Generations of Mankind*. Resolution 43/53, December 6, 1988. New York: UN.

UNGA (2012) *The Future We Want*. Resolution 66/288, September 11, 2012. New York: UN.

UNGA (2014) *The Road to Dignity by 2030: Ending Poverty, Transforming All Lives and Protecting the Planet*. Synthesis Report by the Secretary General on the post 2015 sustainable development agenda. A/69/700. December 4, 2014. New York: UN.

UN Secretary-General's AGECC (2010) *Energy for a Sustainable Future*. New York: UN.

UN Secretary-General Statements (2008) "Opening remarks at joint press conference on climate change Secretary-General Ban Ki-moon, UN Headquarters". http://www.un.org/apps/news/infocus/sgspeeches/statments_full.asp?statID=327 [accessed December 14, 2013].

UN Secretary-General Statements (2013) "Warsaw, 19 November 2013 – Secretary-General's remarks to the Climate Change Conference (UNFCCC COP19/CMP9) High-level Segment".http://www.un.org/sg/statements/index.asp?nid=7290 [accessed December 14, 2013].

UN Secretary-General Statements (2014) "Remarks at the opening of the Plenary of the ASCENT-Abu Dhabi". http://www.un.org/apps/news/infocus/sgspeeches/statments_full.asp?statID=2213#.VCl7eNjwtGE [accessed May 17, 2014].

US Department of Defense (2014) *2014 Climate Change Adaptation Road Map*. Alexandria, VA: Office of the Deputy Under Secretary of Defense for Installations and Environment. http://www.acq.osd.mil/ie/download/CCARprint.pdf [accessed October 13, 2014].

Ürge-Vorsatz, D. and Herrero, T. (2012) Building synergies between climate change mitigation and energy poverty alleviation, *Energy Policy*, 49, 83–90.

van Vliet, O., Krey, V, McCollum, D. *et al.* (2012) Synergies in the Asian energy system: Climate change, energy security, energy access and air pollution. *Energy Economics*, 34(3), S470–S480.

Walsh, J., Wuebbles, D., Hayhoe, K. *et al.* (2014) Our changing climate. In: Melillo, J.M., Richmond, T.C. and Yohe, G.W. (eds) *Climate Change Impacts in the United States: The Third National Climate Assessment*. U.S. Global Change Research Program, pp. 19–67.

WEF (2014a) *Global Risks 2014*, ninth edition. Geneva: World Economic Forum. http://www3.weforum.org/docs/WEF_GlobalRisks_Report_2014.pdf [accessed November 29, 2014].

WEF (2014b) *Climate Adaptation: Seizing the Challenge*. Geneva: World Economic Forum. http://www3.weforum.org/docs/GAC/2014/WEF_GAC_ClimateChange_AdaptationSeizingChallenge_Report_2014.pdf [accessed November 29, 2014].

WHO (2007) *Indoor Air Pollution: National Burden of Disease Estimates*. Geneva: WHO.

WHO (2014a) *Burden of Disease from the Joint Effects of Household and Ambient Air Pollution for 2012*. http://www.who.int/phe/health_topics/outdoorair/databases/AP_jointeffect_BoD_results_March2014.pdf?ua=1 [accessed April 21, 2014].

WHO (2014b) *Burden of Disease from Household Air Pollution for 2012*. http://www. who.int/phe/health_topics/outdoorair/databases/HAP_BoD_results_March2014. pdf?ua=1 [accessed April 21, 2014].

WHO Media Centre (2014) *Household Air Pollution and Health*. Fact Sheet 292. http://www.who.int/mediacentre/factsheets/fs292/en/ [accessed April 22, 2014].

World Bank (2010) *Economics of Adaptation to Climate Change: Synthesis Report*. Washington DC: World Bank. http://climatechange.worldbank.org/sites/default/ files/documents/EACCSynthesisReport.pdf

World Bank (2012) *Turn Down the Heat*. Washington DC: World Bank.

Yoon, S. and Drajem, M. (2014) "China and Indian Leaders Said to Skip UN Climate Summit". *Bloomberg News*, September 4, 2014. http://www.bloomberg.com/ news/2014-09-03/xi-and-modi-said-to-skip-un-climate-summit-later-this-month. html [accessed September 24, 2014].

2

Where is the "Energy" in Global Climate Change Negotiations Outcomes?

Examining Key UN Global Climate Change Outcomes from 1992 to 2014 for References to the Nexus Between Climate Change and Energy Access for the Poor

2.1 Framing the question: Has energy access for the poor been referenced in key outcomes of 20 years of climate change negotiations?

As discussed in Chapter 1, climate change has a known multiplier effect on a range of other socio-economic concerns, and has been documented to have impacts on pressing development issues, including poverty, energy, water, food, agriculture, health, disaster preparedness, habitats, employment and income generation, and socio-economic development in general (World Bank, 2012). Moreover, linkages among climate change, poverty reduction,

Energy and Global Climate Change: Bridging the Sustainable Development Divide,
First Edition. Anilla Cherian.
© 2015 John Wiley & Sons, Ltd. Published 2015 by John Wiley & Sons, Ltd.

and increasing energy access for the poor are not novel in terms of the UN development agenda. The UN's Committee on Development Policy, in its examination of the international development agenda and climate change, stated that "combating climate change and achieving internationally agreed development goals can no longer be placed in separate boxes, but efforts to pursue both objectives should be coherent and mutually reinforcing" (UN Committee for Development Policy, 2007, p. 1).

The broad global consensus that has emerged in recognition of the scope and enormity of the climate change challenge has not been matched in terms of the outcomes of the past two decades of global negotiations aimed at addressing this problem. Global climate change negotiations have been convened on an annual basis from 1992 onwards in all corners of the globe. In addition to these annual, intergovernmental negotiations, a host of other regional, sub-regional, and national technical fora have been held over the past 20 years by innumerable institutions and state actors on a range of related topics. Intergovernmental climate change negotiations have been underway for over two decades, and are still continuing in search for the elusive comprehensive climate mitigation deal. Under the aegis of the United Nations Framework Convention on Climate Change (UNFCCC) to date, 20 consecutive intergovernmental meetings, called Conferences of the Parties (COPs), have been held, starting with COP-1 in Berlin in 1995 and going through to COP-20 in Lima in 2014. From the 2009 COP-15 in Copenhagen to COP-20 in Lima, the most recent COP, the one common denominator for the annual intergovernmental negotiations has been the length and arduousness of negotiations, which typically extend long past the anticipated formal conclusion of the COP to finally arrive at a consensus agreement.

After negotiations that extended more than a day past the official deadline, COP-20's main outcome document – Lima Call for Climate Action (LCCA) – includes a four-page call for action that highlights a more inclusive approach calling on all Parties to submit intended nationally determined contributions (INDCs) in a timely fashion ahead of COP-21. It also includes an Annex containing the elements of a draft negotiating text that is the basis of a more final text expected in May 2015, which in turn is expected to result in a climate change agreement at COP-21 in Paris. What is clear is that INDCs will play a key role in the final outcome of the 2015 COP-21, but what is not clear is whether concrete guidance on energy, in particular the nexus between energy access for the poor and the role of short-lived climate pollutants (SLCPs), will factor into any proposed climate agreement and whether any proposed INDCs will reflect upon the issue of increasing energy access for the poor.

The main objective of this chapter is to review the actual record of key climate change agreed decision outcomes from 1992 to 2014, namely, the UNFCCC, the Kyoto Protocol (KP), the Bali Road Map, the Copenhagen Accord, the Cancun Agreement, the Durban Outcomes, the Doha Climate

Gateway, the Warsaw Outcomes and the 2014 LCCA, to see what specific references and policy linkages, if any, have been made to energy access for the poor and/or poverty reduction issues. These agreed global outcome documents constitute the primary source material for this chapter, and they will be examined for their concrete references and linkages to "energy access" and "poverty eradication/reduction". Based on an examination of these crucial climate change outcomes, this chapter will demonstrate that concrete references to "energy access" or "access to energy" are missing, and that even broader references on energy and poverty reduction are minimal or lacking in terms of providing policy and programmatic linkages and guidance within the climate change context. To the extent that is relevant, the chapter will extend the argument that the nexus between energy access for the poor and climate change is germane to developing countries, particularly the large aggregate greenhouse gas (GHG)-emitting developing countries that play a critical role in any future resolution of global climate change and energy for sustainable development negotiations.

The explicit articulation of the precautionary principle within the UNFCCC should have bolstered intergovernmental action on global climate change, but instead an examination of the actual globally agreed outcomes of the 20-year period of intergovernmental climate change negotiations reveals a somewhat different trajectory of action. As this chapter demonstrates, the examination of key outcomes of intergovernmental climate change negotiations reveals a puzzling disconnect between the global rhetoric on climate change and the actual adoption of concrete precautionary measures, particularly those related to energy access for the poor. The examination of key agreed global climate change outcomes reveals that references to energy access for the poor are entirely absent, and that concrete programmatic guidance on the linkages between energy and poverty reduction are largely missing.

The chapter does not seek to assign responsibility for the lack of systematic and sustained linkages among energy sector mitigation, energy access and poverty reduction in key global climate change outcomes. Nor does it aim to provide a detailed summary or historical accounting of the more than 20-year process of intergovernmental climate change negotiations in terms of institutional, national or global actors or the climate change regime as a whole. Any such review or summary of the negotiations process could by default be found lacking, given the complexity of the topic, the timescale, and the number of actors involved. In other words, any historical review or evaluation of the negotiations from the perspective of a set of actors or institutions would be somewhat subjective because a truly comprehensive review would need to capture an extremely wide array of diverse views and represent a range of different country-driven and/or institutional actors and positions over the 20-year time span. Accordingly, any such analytic summary of the 20 plus years of climate change negotiations is beyond the purview of this current book. Furthermore, this chapter is not intended to provide an insight into how the climate change negotiations got to the state they are currently in. Given the lack of overall information about the apparatus and process for climate change policymaking in developing countries, it would be useful to undertake detailed examinations of the same, particularly within large developing countries that have been active in the negotiations over the course of the 20-year period. However, such examinations require unfettered and equitable access to the policymaking apparatus of relevant developing countries over a comparable time frame, which is hard to access, nevertheless worthwhile to aim for.

The chapter begins with a brief review of the analyses focused on understanding and explaining the global climate change negotiations from a variety of perspectives. The brief overview of existing analyses of climate change negotiations should not be construed as exhaustive or complete, and is provided only as a contextual background for those interested in seeing the broad trends of climate change negotiation analyses over course of 20 years. The main focus of the chapter is to examine key globally agreed outcomes emanating from the UN-led climate change negotiations process ranging from the 1992 UNFCCC to the 2013 Warsaw outcomes, and the more recent 2014 LCCA for references and linkages to energy access and poverty reduction. The concluding segment of the chapter summarizes the findings, which point to a lack of references to energy access for the poor in key climate change outputs. The lack of concrete references on energy access for the poor within the key global climate change outcomes is highlighted as a concern in the current UN quest for a "universal" and integrated post-2015 development agenda.

2.2 Global climate change negotiations analyses: A brief overview of broad trends

Global climate change negotiations initiated from 1992 onwards have been occurring with greater periodicity and frequency, but unfortunately have not yet resulted in a comprehensive consensus-based climate change mitigation agreement. The evolution and the slow pace of global environmental governance and negotiations, including climate change negotiations, have been the subject of considerable analyses. Focused on the prospects of advancing negotiations on the concept of the green economy at the 2012 Rio+20 Summit, Haas (2012) has referenced the slow pace of the negotiations and called for more clarity so as to allow for clearly defined political effort and rules and regulations that would permit robust investment in new technologies, while Chasek and Wagner (2012) provide perhaps the most comprehensive historical analyses and insider-based perspectives of the 20 years of multilateral environmental negotiations. But, in the case of global climate change, the pace has been particularly fractured and fractious. Countries have so dragged their feet in combating climate change that the risk of severe economic disruption has risen, and "another 15 years of failure to limit carbon emissions could make the problem virtually impossible to solve" (Gillis, 2014).

There were 11 intergovernmental negotiating committee meetings in the lead-up to the historic 1992 UNFCCC, and there have been 20 COPs to the UNFCCC to date, and all of them have included tense exchanges of views between developed countries and developing countries, which have grown especially acrimonious in more recent years as the issue of developing countries taking on responsibility to mitigate against climate change has come to the fore. Climate change negotiations have long been characterized by political differences between developed (Annex I Parties to the UNFCCC) and developing countries (non-Annex I Parties) over who bears responsibility for causing the problem, and assessing the scope, scale, and extent of this responsibility, and how, when, and by what means to provide financial resources, technological inputs and/or institutional capacity building that can allow developing countries to adapt and respond to the problem.

Patterson and Grubb argued back in 1992 that, of all the global environmental challenges emerging from the historic 1992 Rio UN Conference on Environment and Development (UNCED) process, climate change would be the "acid test" of whether countries would seriously commit to responding to addressing global environmental challenges, because the response to limit climate change would "go to the heart of their political and industrial structure". They also noted that climate change became a major political issue as a result of a series of scientific conferences and also because 1988 was a year of freak weather (Patterson and Grubb, 1992, p. 293). More than 25 years later, after more than a decade of freak and extreme weather events, ranging from superstorms to huge forest fires and droughts, the question as to exactly when, whether, and how countries will commit to climate change mitigation remains unanswered. Meanwhile, as far back as 1997, in writing about the contradiction associated with the role of the World Bank, as the biggest funder of fossil fuel projects, asking for a leading role in the KP's push to reduce global carbon emissions, Flavin (1997) pointed out that although the KP covered only developed countries, developing countries' emissions had grown by nearly 75%, powered mainly by fossil fuels and, as a result, countries like China, India, and Brazil are becoming increasingly important to the stability of the global climate (p. 26).

The 20-year cycle of intergovernmental climate change negotiations has been characterized as stalemated by a growing impasse between Annex I Parties and non-Annex I Parties to the UNFCCC – essentially the divide between developed and developing countries – between those who have taken on a historical responsibility for anthropogenic GHG emissions and those who have not (Gupta, 2000). But this divide has become more of a static representation and does not really capture the more recent, dynamic reflection of the growing role of large aggregate GHG-emitting developing countries – Brazil, South Africa, India, and China (the BASIC countries) – and newer and more vocal groupings of countries. In fact there are a large number of different groupings of countries based a variety of criteria ranging from regional representation to interests, with some countries belonging to more than one group. Based on the long-standing tradition of the UN, Parties to the UNFCCC are organized into five regional groups, which form the basis for electing the Bureau membership to the Convention and its Protocol's various bodies: African States, Eastern European States, Latin American and Caribbean States, and the Western European and Other States (which includes Australia, Canada, Iceland, New Zealand, Norway, Switzerland, and the United States of America). The 28-member European Union is perhaps the most regionally integrated group that has agreed common negotiations positions, but 134 developing countries currently use the forum of the Group of 77 and China as a basis for shared negotiating positions as well. The G-77 and China, which represents a large majority of UN member countries, comprises a very diverse range of countries with widely varying interests on climate change. As a result of the variations amongst countries regarding the climate change challenge, a dizzying array of further sub-grouping of countries are used by Parties to negotiate including:

- 48 countries defined by the UN as the Least Developed Countries (LDCs);
- 40 nation coalition comprising the Small Island Developing States (SIDS);
- 44 island and coastal countries that comprise the Alliance of Small Island States (AOSIS);
- Organization of Petroleum Exporting Countries (OPEC);
- Umbrella Group which is a less defined coalition of non-EU members made up of Australia, Canada, Japan, New Zealand, Kazakhstan, Norway, the Russian Federation, and the US;
- Environmental Integrity Group comprising Mexico, Liechtenstein, Monaco, the Republic of Korea and Switzerland;
- Bolivarian Alliance for the Peoples of our America (ALBA in Spanish);
- Like-Minded Developing Countries (LMDCs), which includes, amongst other countries, China, India and Saudi Arabia;
- Central Asia, Caucasus, Albania and Moldova (CACAM);
- Central American Integration System (SICA in Spanish);
- Association of Independent Latin American and Caribbean States (AILAC).

There is a considerable body of literature that has focused on analyzing the global climate change negotiations process through its various outcomes and stalemates over the course of the over 20-year period since the adoption of the UNFCCC. A variety of factors, ranging from weak enforcement norms to the role of non-state actors (i.e., civil society and non-governmental organizations [NGOs]) have been highlighted by diverse authors as a means of explaining why the global climate change negotiations are in the state they are in. The bulk of the analyses on the climate change negotiations has focused on examining the role of the climate change regime, including the rule of norms and procedures, and/or key actors, be they state or non-state, for their role in shaping/influencing a particular set of climate change negotiations or outcome (Levy *et al.*, 1993; Bernauer, 1995; Keohane and Levy, 1996; Bodansky and Diringer, 2010).

Studies in environmental politics and climate change policymaking have tended to focus on understanding the environment–development linkages in terms of political, socio-economic concerns, and institutional perspectives existing in or pertinent to developed countries. It is possible to identify three broad categories or models of environmental policymaking in this regard. The earliest of these and the one most commonly associated with European and US environmental policymaking is the electoral/political pressure model, which states that public concern for the environment creates electoral or political pressure or environmental coalitions/parties that influence outcomes and result in particular sets of legislation or stances (Downs, 1972). A second model that can be identified is the bureaucratic or organizational change model, which argues that structural changes, in particular in organization or state bureaucracy, lead to a particular set of environmental policies or issues being identified or championed (Lundqvist, 1980). A third model is the exogenous change model, which argues that external pressures or norms generated by global institutions and international regimes contribute to changes in national policymaking and stances (Litfin, 1993; Levy *et al.*, 1994).

What is interesting to note, however, is that despite two decades of global climate change negotiations, there remains a large gap in analyses focused on understanding and explaining the diverse perspectives of developing countries, as most of the focus on climate change negotiations has tended to be directed towards the role of global institutions, or state and non-state actors relevant to developed countries. This has been changing somewhat in the past decade as increasing attention has been paid to the role of BASIC countries as large aggregate GHG emitters, but still the complex processes by which developing countries balance policy demands related to climate change and energy security and poverty reductions remains largely unexplored.

An extensive body of literature has coalesced around the issue of how best to resolve and/or smooth out perceived or existing logjams or blocks in the global climate change negotiations apparatus and process. Much of this analysis has been conducted by research and global institutions focusing on highlighting policy options for the consideration of climate change policymakers and negotiators in the lead-up to specific COPs or in providing post-mortem analyses following the outcomes of a particular COP (Bodansky, 2011). For instance, the failures and proceedings of the sixth COP (COP-6), held in the Hague in 2000, and the 2009 Copenhagen COP have been subjected to considerable political and analytical scrutiny. In their article, published in a special issue of *International Affairs* focused on climate change, Grubb and Yamin (2001; see also Ott, 2001) provide what they call an "initial post-mortem", where they note the confusion that occurred during the close of the Hague COP, where an 11th-hour attempt at a bilateral EU–US deal was made outside the main negotiations by the UK, which turned out not to have the mandate from the EU to forge any such deal. This in turn caused developing country negotiators to express their frustrations at being shut out of negotiations (Grubb and Yamin, 2001, p. 263).

The role of climate governance mechanisms and arrangements has also been focused within the context of the current international climate change regime in terms of state and non-state actors, including non-governmental and intergovernmental entities and processes. Sometimes, such analysis of intergovernmental climate negotiations has looked at both state and a wide variety of non-state actors in the same frame. For instance, Sjöstedt and Penetrante (2013) identified "road blocks" to the KP from five professional perspectives – a top policymaker, a senior negotiator, a leading scientist, an international lawyer, and a sociologist – and also listed major problems including the role of great power strategies, leadership, NGOs, capacity and knowledge building, airline industry emissions, insurance and risk transfer instruments, problems of cost–benefit analysis, and the Intergovernmental Panel on Climate Change (IPCC) in the post-Kyoto era. A somewhat different analytic approach has been the game theory approach, where, for instance, game theory methodology is use to predict or analyze how particular groupings of countries may or may not behave. Using five-stage sequential game methodology, Císcar and Soria (2002) tried to replicate the Kyoto and post-Kyoto scenarios for Annex B (developed) and non-Annex B (developing) countries.

In terms of assessing the future scope and scale of climate change responses and progress in negotiations, there has been a good deal of focus on issues of equity and fairness, which are issues that resonate with the negotiating policy stances of developing countries. For example, Rosa and Munasinghe (2002) pointed out that since climate change is an issue in which all human beings are potential stakeholders, equity and ethical considerations such as morality, rights and law, social attitude, and beliefs have an important role to play in understanding and enhancing the global climate change negotiations. The issue of fairness and equity has also been considered by Soltau (2009), who examined the scientific evidence and current state of play in the climate change regime and outlined elements of a working consensus on the concept of fairness that could be used to assign climate change responsibility. Roberts and Parks (2007), meanwhile, analyze the role that inequality plays to develop new measures of climate-related inequality and analyze patterns of "emissions inequality" to argue that the current negotiating gridlock between north and south will remain unresolved in the absence of a global bargain on environment and development. In contrast to the more abstract examination of the role of equity and fairness to explain the status and/or nature of the climate change negotiations, other analysts have focused on more concrete aspects, such as the role of specific organizational processes and actors. For Depledge (2005), the frame of reference in terms of analyzing the climate negotiations is not equity but the actual organization of the negotiations process. For her, factors such as the role of the Chair, choice of negotiating arenas, and texts are taken for granted but are crucial in understanding complex global negotiations on climate change, which involve large numbers of countries and the need to address enormously complex issues with conflicting agendas.

An additional important category of analyses that focuses on the role of non-state actors in influencing both global and national environment stances has focused on what Haas referred to as the role of epistemic communities, which can broadly be characterized as a transnational network of professionals with recognized expertise in a particular domain, an authoritative claim to policy knowledge within their domain, and shared common views about the causes and solutions of a policy problem (Haas, 1992). Over 20 years ago, Adler and Haas (1992) outlined a four-step process by which epistemic communities engage in policy evolution, including policy innovation, diffusion, selection, and persistence. Given the important role of scientific coalitions and consensus and the role of policy experts in explaining the emergence and evolution of climate change as a global challenge and the negotiations associated with addressing this issue, the tracing of epistemic communities and their roles within and across developing countries remains a useful issue to explore. Individual developing countries or institutions and actors based in these countries played an influential role in the course and evolution of the global climate change negotiations from 1992 onwards. The roles of key negotiators, including Ambassadors Chandrashekar Dasgupta and Sreenivasan of India, who were extremely active in shaping the climate change debate in the early years, Ambassador Raul Estrada Oyuela of Argentina, credited with shepherding the KP and serving as chair of the UNFCCC Subsidiary Body on Implementation (SBI), Ambassador John Ashe of Antigua and Barbuda who was the first chair of the Protocol's Clean Development Mechanism, chair of the SBI, and the chair of the Ad Hoc Working Group on the KP, as well as many other negotiators from a range of developing countries, are worth examining.

The socio-economic challenges facing many developing countries do not and did not mean that developing policymakers were somehow not interested and active participants in the negotiations leading up to and after the UNFCCC. As Chapter 3 demonstrates, developing countries' negotiators were also active in the other UN-led processes related to environment and development in the 1990s onwards, which was reflective of the new trend towards the internationalization or globalization of environmental problems (Hurrell and Kingsbury, 1992, p. 2). It would be mistake to assume that environmental issues are somehow largely the purview of policymakers in developed countries, and that citizens of developing countries are somehow not as interested in environmental concerns. In fact, 20 years ago, a review of the coverage by leading national newspapers of the UNCED in India indicated that over 280 articles on the proceedings were published from March to July 1992 (TERI, 1992).

From the perspective of developing countries, a considerable amount of discussion within the context of the climate change negotiations has been to highlight the adverse impacts of climate change and call for financial, capacity, and technical resources and inputs that are needed if vulnerable countries and communities are to deal with adaptation. Policymakers from vulnerable least developed countries (LDCs) with large coastlines and small island developing states (SIDS) – viewed as frontline states facing climatic threats – have long acknowledged that global climate change is one of the most serious challenges confronting development and have chosen to be vocal advocates of climate change mitigation (Quarless, 2007). Issues like sea-level rise, coral bleaching, increases in the intensity and frequency of hurricanes, elevated coastal water temperatures, and coastal zone inundation threaten their endemic and fragile biodiversity resources and their overall future. What is especially troubling for SIDS and LDCs, for instance, is that they contribute the least – in terms of per-capita and aggregate GHG emissions – to global climate change but happen to be the most vulnerable to adverse climatic impacts. Even though climate change mitigation options pursued by these countries appears negligible in terms of the global aggregate, many of these countries have demonstrated a keen interest in the development and use of renewable energy (Cherian, 2007).

For many of these countries, the key to climate change resides in the extent and efficacy of implementing adaptation policies and measures. Khan, for example, argues that the legal framework for adaptation should be strengthened and that planned policies and measures for reducing vulnerability to the adverse impacts of climate change should be framed as a global responsibility within a legally binding regime along the lines as climate change mitigation (Khan, 2014). This is clearly an important issue from a development perspective, but while recognizing the crucial role of adaptation in development and poverty reduction efforts, the immediate focus of this chapter is not on addressing adaptation issues. More comprehensive and detailed studies of the policy apparatus and frameworks by which developing countries formulate policy and negotiations stances on global climate change need to be undertaken in order to better understand the energy–global–climate change nexus.

In terms of climate change mitigation analyses, it is important to note that options for differentiating actions or commitments have been considered over a long period of time and are not issues that are just now coming to the fore. Authors like Grubb and Anderson (1995) highlighted specific factors and options that are critical to any future climate change regime, including the role of differentiated commitments by countries. During the context of the climate change negotiations initiated in Berlin in 1995, which would eventually lead up to the KP, there were calls for new commitments to be differentiated between different kinds of industrialized countries as well as heated discussion on the role of development country involvement. The issue of differentiation of commitments was for instance, discussed at an influential workshop convened by the Royal Institute of International Affairs, held in 1996, which brought together leading negotiations and analysts and included discussions on ways of grouping national commitments as well as proposals for indexing emission targets and sectoral commitments (Paterson and Grubb, 1996).

Ever since the debate over commitments began in the context of climate change negotiations over the KP, there has been a pitched tension between countries, which has only become more elevated with the growing aggregate GHG contributions made by BASIC countries. Fundamental to this debate is the frequent invocation of the principle of "common but differentiated responsibilities" between developed and developing countries that has been raised over the course of the past two decades, particularly by developing country negotiators. This principle has a particular resonance when seen in relation to the linkage between national energy needs and issues and global climate change responsibilities. Given the differential yet direct and influential role of energy in driving economic development and global climate change in all countries, the linkages between the energy sector and global climate change have proved extremely difficult to address from a political and socioeconomic perspective at national and global levels. What is fundamentally at issue is who has control, and how, over the global space for economic development fueled by the growing energy sector and its attendant GHG emissions.

The relationship between energy and anthropogenic climate change has been viewed by all countries as crucial for economic development. But the specific linkages among lack of energy access, poverty reduction, and adverse impacts of climate change have tended not to be addressed by many analysts of climate change negotiations, who focus primarily on analyzing the failures and stalemated status of global climate change negotiations. Rather than addressing the development challenges, including poverty reduction and lack of energy access facing developing countries, for instance, Victor (2008) claims that "the most important yet challenging aspect of international climate policy" has been to encourage developing countries to take on climate change mitigation but that these countries have refused to undertake "credible" measures to reduce their growing GHG emissions. According to Victor (2008), the two reasons for this reluctance by developing countries is that they "put a higher priority on economic growth far above distant, global environmental goods" and, secondly, "the governments of the largest of these countries, namely China and India – actually have little administrative ability to control emissions in many sectors of their economy" (pp. 3–4). Victor argues that the task for the so-called "enthusiastic nations" is to craft deals – referred to as "Climate Accession Deals (CADs)–borrowing concepts from the General Agreement on Tariffs and Trade/World Trade Organization (GATT/WTO), the International Monetary Fund (IMF), the Organisation for Economic Co-operation and Development (OECD) and other international economic regimes that have confronted the thorny issue of differential interests and capabilities (Victor, 2008, p. 37). But the idea that only developing countries are preoccupied with short-term economic development compared with longer-term environmental cost–benefit analyses done by developed countries sets up a false dichotomy, one that is contradicted by statements made by China and South Africa at the 2014 Climate Summit. Moreover, Victor's characterization of large aggregate GHG-emitting developing countries like China and India as "reluctant" to engage in climate mitigation versus the more euphemistically titled "enthusiasts" (mainly the industrialized world), who, according to Victor, can attempt to tame GHG emissions growth in developing countries not through previously used carrots and sticks (trade sanctions) but through new CADs, is troubling. His argument that so-called climate "enthusiasts" can somehow coax so-called reluctant nations (mainly the developing world) into a common global effort flies in the face of the fact that the so-called reluctants (namely, BASIC countries) have, in reality, made advances in curbing GHG emissions on their own volition, including through the use of renewable energy and energy efficiency targets, and that attempts to coax appear to harken back to a different historical era.

While there is no doubt that climate change negotiations have been gridlocked, some analysts, like Hoffman, have seen in the gridlock new opportunities for multilateral non-UN-driven climate partnerships. Here the argument is that although UN-driven global treaty making is stalemated in terms of the KP negotiations, there have been diverse experiments responding to climate change, such as the Asia-Pacific Partnership for Clean Development and Climate Change and the Major Economies Forum on Energy and Climate, that have arisen as a response (Hoffman, 2011). This idea that climate change does not need to be addressed solely through the mechanism of a global UN-led comprehensive treaty is an interesting one, borne out by the stymied pace of climate change negotiations, which can be contrasted with the new innovative and collaborative actions happening outside of the UN by cities, state governments, and civil society actors, and some of these will be discussed in Chapter 4.

There is a general scarcity of analyses on the role of large developing countries that actively negotiate climate change and energy. Ravindranath and Sayathe (2002) made an early effort to address this knowledge gap, but there needs to be a lot more work focused in this area. More recently, there have been analyses that have broken from the traditional characterization of a recalcitrant south (developing countries) facing off an eager north (developed countries). For instance, Mathoo and Subramanian (2013) argued that it is time to change the "narrative" of mutual recrimination between rich and poor countries to one based on shared interests, and that global leaders need to shift attention from emission cuts to technology generation, and ditch the "cash-for-cuts" approach and replace it with one that requires contributions that are calibrated in scope and scale to national development levels. In their analysis of the 2012 Doha COP, Roberts and Edwards (2012) argued that it is crucial to move away from the media fascination to report on negotiations conflict between the global North and the South. They point out that it the "revolt of the middle" – that is, neither the poorest nor the richest – which is characterized by the creation of the 2012 negotiating bloc AILAC that represents a new shift, with these developing countries deciding to launch plans for low-carbon development without waiting for action from developed, richer countries. After 20 years of negotiations with diminishing returns in terms of a comprehensive framework agreement to date, these arguments appear persuasive.

2.3   Examining key global climate change outcomes for references to the "energy" and "poverty reduction" nexus: Has "energy access for the poor" been referenced in key agreed outcomes?

The discussion below is focused on examining key globally agreed outcomes/decisions resulting from the intergovernmental climate change negotiations, to see whether and how energy access and poverty reduction are referenced. The scope and duration of global climate change negotiations are an indication that there has been no dearth of negotiating opportunities and fora. In fact, intergovernmental climate change negotiations are only one part of the global climate change complex, which comprise of wide variety of technical and policy meetings convened by diverse actors, ranging from governments and institutions to civil society actors. As noted previously, this section will not provide analyses as to which actors or sets of actors were or were not responsible for any particular references and linkages on energy and poverty reduction being included or excluded; nor will it provide detailed historical assessments of what transpired during the course of the 20-year journey of climate change negotiations. The aim is not to provide a timeline or detailed accounting of the climate change negotiations or to comment on or provide input into all the various procedures and debates that have occurred in the two decades of global negotiations, but to provide a context for the key climate change outcome decisions being examined.

2.3.1   The early years: Examining the 1992 UNFCC and the 1997 KP for references to the energy and poverty reduction nexus, and to energy access for the poor

Starting in the late 1980s, a convergence of growing scientific and environmental awareness over the impacts of human activities on the global environment led to a growth of scientific and policy interest in global climate change. While climatology and the study of the greenhouse effect stretch back in time to the past century (Houghton, 2009), organized and intergovernmental scientific interest in the anthropogenic impacts of global climate change may be traced back to convening of the First World Climate Conference held by the World Meteorological Organization (WMO) in 1979. The Conference led to the creation of the World Climate Programme (WCP), which was the first global attempt to conduct research on the human-induced – anthropogenic – causes and effects of global climate change, and it was followed by the World Commission on Environment and Development (WCED) (discussed at length in Chapter 3), which harnessed attention to climate change as an issue of global significance.

The period from 1988 to 1990 was crucial in terms of bringing global prominence to the issue of climate change. Responding to a growing concern that anthropogenic increases of GHG emission aggravated the natural greenhouse effect, the WMO and UNEP established the IPCC in 1988. Over the years, a voluminous body of scientific literature compiled by the IPCC has focused on various aspects of what is currently referred to as the global climate change problem, including the range of GHGs and their varied impacts. Also in 1988, the World Conference on the Changing Atmosphere: Implications for Global Security – more simply named the Toronto Conference, given its venue – was an early precursor in the pattern of convening scientists and policymakers around the issue of climate change and issuing a declarative conference statement. The Toronto Conference brought together more than 300 scientists and policymakers and issued a final statement that called for support for the work of the proposed IPCC and the future development of framework protocol for the protection of the atmosphere (WMO, 1989).

In terms of intergovernmental discussion within the context of the UN, global climate change first emerged on the intergovernmental arena as a global concern in 1988, when the UN General Assembly (UNGA) adopted a resolution sponsored by the Government of Malta, recognizing climate change as a "common concern of mankind" (UNGA, 1988). This initial resolution on global climate change was immediately followed by a series of other UNGA resolutions that emphasized developing countries' concerns over climate change, and focused on the specific adverse effects of sea-level rise on islands and low-lying coastal states. (UNGA, 1989a,b).

There were several important milestones that occurred in 1990 which would have a direct impact on the intergovernmental negotiations process that would culminate in the adoption of the 1992 UNFCCC. In September 1990, WMO and UNEP established an Ad Hoc Working Group of Governmental Representatives to discuss the structure and format of the global climate change negotiations. The group proposed that all international climate change negotiations take place under a single forum within the UN. Then in late December 1990, the UN passed a historic Resolution 45/212 that established the International Negotiating Committee (INC) and charged the INC with responsibility for preparing a Framework Convention on Climate Change. The resolution was co-sponsored by 58 nations and spearheaded by Ambassador Alexander Borg Olivier, Malta's Permanent Representative to the UN. In a personal interview conducted in September 1993, Ambassador Borg Olivier, explained that he reviewed a variety of emerging environmental concerns, and personally pushed for Malta to take a lead role in sponsoring this resolution, given the potential threat that global climate change could have for future generations. Malta would also provide the first and long-serving Executive Secretary of the UNFCCC – Michael Zammit Cutajar.

According to the text of the resolution, the INC was to provide the sole forum in which an intergovernmental negotiating process on climate change would proceed. The resolution envisaged the administration of the INCs under the direct auspices of the UNGA, with the assistance of the WMO and UNEP. The UNGA created a dedicated Trust Fund to cover the costs of the INCs, as well as a Special Voluntary Fund to enable the participation of developing countries' representatives (UN, 1990). Funding for the participation of developing country representatives continues to be a key issue to this day, as this Fund continues to provide finance for the participation of at least one expert representative from the LDCs and SIDS in the global climate change meetings, which in time would increase in frequency and in terms of being convened in diverse locations.

The 1990 Second World Climate Conference was an early yet significant step towards the emergence of the UNFCCC, particularly from the immediate perspective of developing countries' role in climate change. Sponsored by the WMO, UNEP and others, the 1990 Second World Climate Conference was an early yet significant step towards the emergence of the

UNFCCC, and its objectives were to review existing reports and recommend policy actions. Supported by UNEP and WMO, the mandate of the INC/ FCCC was to prepare an effective framework convention on climate change. From the 1990s onwards, the global debate hinged on determining the nature, scope of causal impacts and responses to increases of GHGs in the Earth's atmosphere, with the principal challenges being whether and how to attribute causation and assign responsibility for the Earth's changing climate – issues that remain largely unresolved in terms of the global comprehensive framework for concrete mitigation targets that can be adopted by all countries even today.

On December 11, 1990, the 45th session of the UN General Assembly adopted a resolution that established the Intergovernmental Negotiating Committee for a Framework Convention on Climate Change (INC/FCCC). The INCs, which included representatives from over 150 states, held five sessions between February 1991 and May 1992 prior to adoption of the Convention. Hyder (1994) has argued effectively that the decision to locate the climate change negotiations within the UN framework, and the move towards ensuring the universal participation of countries within the INC format, was welcomed by developing countries because it provided an opportunity for broader scale participation in the climate change negotiations (Hyder, 1994, p. 203) The negotiations in these INCs included spirited exchanges between developing and developed countries, but also included tensions between oil-producing developing countries and small island nations comprising the AOSIS, which are vocal advocates for climate action as they saw themselves most vulnerable to the adverse effects of climate change. These INCs were early precursors of what would become increasingly contentious negotiations over issues such as binding commitments, targets, and timetables for the reduction of carbon dioxide emissions, financial mechanisms, technology transfer, and "common but differentiated" responsibilities of developed and developing countries, to name but a few.

UNFCCC   At the global level, the most definitive outcome of intergovernmental negotiations on climate change is the UNFCCC. In the ensuing 20 years since it was adopted, it remains the authoritative outcome of global climate change negotiations, and there has been no other comprehensive framework document that has been agreed to. It was one of the three principal conventions adopted at the historic UNCED – the Earth Summit. It was adopted on 9 May 1992, and opened for signature at the UNCED in June 1992 in Rio de Janeiro, Brazil. The Convention entered into force on March 21, 1994 (90 days after receipt of the 50th ratification), and since then, 194 countries (and one regional economic integration organization) have signed on as Parties to the UNFCCC representing universal membership of the UN.

After the adoption of the Convention, the INC met six more times, but the key negotiating stumbling blocks remained, including how best to address commitments; methodologies to estimate the removal of carbon dioxide by "sinks", namely forests and oceans; arrangements for technical and financial support to developing countries; procedural and legal matters; and institutional matters. What is important to note in terms of linkages between energy and global climate change mitigation is that the UNFCCC's ultimate objective of achieving the stabilization of atmospheric GHG concentrations does not include any specific legally binding targets or timetables for energy-related action that would either commit or guide countries as to exactly how this stabilization would occur, nor does it reference in any precise or concrete manner what it qualifies as "dangerous anthropogenic interference". Interestingly, the UNFCCC notes only that "such a level" should be achieved "within a time-frame sufficient" to allow for the natural adaptation of eco-systems, and to ensure that food production is unthreatened and economic development can proceed sustainably (Information Unit on Climate Change [IUCC], 1992, p. 5).

Based on interviews with a variety of key lead climate change negotiators from developing countries, in the lead-up to and the subsequent adoption of the UNFCCC, two key criteria were explicitly stated and were seen as relevant:

- The role of historical responsibility of developed countries (Annex I Parties) in taking on mitigation actions or commitments.
- The recognition that the economic development trajectories of developing countries could not be constrained, in particular Article 3.2 and Article 4.7 of the UNFCCC (Cherian, 1997).

The record of global climate change negotiations reveals that developing country negotiators from diverse sets of countries, including large ones like India, China, Brazil, and Argentina, and smaller ones, from the AOSIS, like Vanuatu and Antigua and Barbuda, played an active role in the climate change negotiations, contradicting the idea put forward at the time by Hawkins (1993) that "future global climate change has been a low priority for most developing countries" (p. 233). Based on inter-views with several lead climate change negotiators from India, Brazil, and AOSIS, energy-related development needs were seen as a key frame of reference in the global climate change negotiations leading up to UNFCCC (Cherian, 1997).

Interestingly, a review of the UNFCCC reveals that the entire text of the Convention contains only two explicit references to "poverty". The phrase used twice in the UNFCCC is "poverty eradication" rather than "poverty reduction". The first reference to poverty eradication is contained in

Article 4.7 and is particularly relevant from the perspective of understanding the role of developing countries in the implementation of the UNFCCC, because it stated that:

> The extent to which developing country Parties will effectively implement their commitments under the Convention will depend on the effective implementation of developed country Parties of their commitments under the Convention related to financial resources and transfer of technology and will take fully into account that economic and social development and poverty eradication are the first and overriding priorities of developing country Parties. (IUCC, 1992, p. 11)

This remains an explicit reference to the primacy accorded to poverty eradication by developing countries within the UNFCCC. Additionally, there was a reference to the "eradication of poverty" in the preamble to the UNFCCC, which affirmed that climate change responses should, *inter alia*, take into "full account the legitimate priority needs of developing countries for the achievement of sustained economic growth and the eradication of poverty" (IUCC, 1992, p. 4). The UNFCCC contains no references to the linkages between eradication of poverty and the role of increasing access to energy services for the poor.

Despite the central role of energy in global climate change mitigation, the final agreed text of the UNFCCC contains no references to the term "access to energy" or "energy access" and only six references in total to the term "energy". Two of these six references to energy are in the final paragraph of the preamble of the UNFCCC, which recognizes that

> All countries, especially developing countries, need access to resources required to achieve sustainable social and economic development, and that in order for developing countries to progress towards that goal, their energy consumption will need to grow taking into account the possibilities for achieving greater energy efficiency and for controlling greenhouse gas emissions in general, including through the application of new technologies on terms which make such an application economically and socially beneficial. (IUCC, 1992 p. 4)

This was the first clear reference to the idea that developing countries need to have access to energy resources that would allow them to achieve sustainable socio-economic development, but here the issue of increasing energy access to reduce poverty and improve sustainable human development is not explicitly mentioned. The reference to poverty eradication being a priority need for developing countries was contained in the previous paragraph of the preamble, and kept distinct from the concept that energy consumption of developing countries will need to grow.

There is **no** stand-alone paragraph referencing the energy sector within the UNFCC. The only reference to the energy sector and the need to address climate change is found in Article 4.1 (Commitments), which calls on Parties, based on their "common but differentiated responsibilities" and "specific national development priorities, objectives and circumstances", to:

> ...promote and cooperate in the development, application, diffusion and transfer, of technologies, practices and processes that control, reduce or prevent anthropogenic emissions of GHGs not controlled by the Montreal Protocol in all relevant sectors, including the energy, transport, industry, agriculture, forestry and waste management sectors". (IUCC, 1992, p. 7)

This call to all Parties to work together to control, reduce, or prevent GHGs in the energy sector is critical to climate change mitigation, but it is important to note that the energy sector which has cross-sectoral impacts on transport, industry, agriculture, and waste is not given any predominance over the other sectors. In spite of its critical role in contributing to anthropogenic GHG emissions, there is no concrete guidance provided as to specific measures to control, reduce, and prevent GHGs in the energy sector or how the promotion and cooperation will occur.

Two of the three remaining references to energy are not related to climate change mitigation or sustainable energy but instead reference a completely different issue. The reference in Article 4.8 (h) focuses on the needs of countries "whose economies are dependent on income generated from the production, processing, export and/or consumption of fossil fuels and associated energy intensive products" (IUCC, 1992, p. 11). The second reference contained in Article 4.10 of the UNFCCC, calls on Parties to take into consideration the "needs of developing country Parties, with economies that are vulnerable to the adverse impacts of the implementation of measures to respond to climate change" – and here the explicit reference is again to "Parties with economies that are highly dependent on the production, processing, export and/or consumption of fossil fuels and associated energy- intensive products" (IUCC, 1992, p. 11).

In other words, both the references to energy in this context relate to countries whose economies will be negatively impacted if climate change mitigation were to occur and not to any specific guidance on energy sector-related mitigation. The final and sixth reference to energy is contained in Article 7.5, which lists a series of UN agencies including the International Atomic Energy Agency.

Kyoto Protocol (phase 1)   After a series of late-night negotiations, countries adopted the Kyoto Protocol to the UNFCCC in 1997. The KP's main objective was to limit emissions of certain GHGs not controlled by the Montreal Protocol. The KP committed Parties (referred to as Annex I Parties) to individual, legally binding targets to limit their GHG emissions. Only Parties to the Convention that have also become Parties to the Protocol (i.e., by ratifying, accepting, approving, or acceding to it) are bound by the Protocol's commitments. The KP had 83 signatories, and currently includes 192 Parties (191 States and one regional economic integration organization) – it does not cover the full spectrum of Parties that have signed on to the UNFCCC, and includes some notable exceptions, namely the United States and, more recently, Canada, that ratified the KP initially but withdrew from the KP in December 2012 just days prior to the negotiations related to the KP's second commitment period.

Within the KP, 37 industrialized countries and the European Community have committed to reducing their emissions by an average of 5% by 2012 against 1990 levels. Individual targets for Annex I Parties are listed in the KP's Annex B. These add up to a total cut in GHG emissions of at least 5% from 1990 levels in the commitment period 2008–2012. Under the KP, Parties with obligations (i.e., those in Annex B of the KP, primarily developed countries) must first and foremost take domestic action against climate change. However, the Protocol also allows them to meet their emission reduction commitments abroad through three market-based mechanisms, namely, emissions trading (known as "the carbon market"); the clean development mechanism (CDM); and joint implementation. The mechanisms help stimulate green investment and help Parties meet their emission targets in a cost-effective way.

Article 12 of the KP establishes a CDM which permits Annex B Parties to implement emission-reduction projects in developing countries. Such projects can earn saleable certified emission reduction (CER) credits, each equivalent to 1 tonne of CO, which can be counted towards meeting Kyoto targets. The CDM is the first global, environmental investment and credit scheme of its kind, providing standardized emissions offset instruments – CERs. The mechanism is seen by some as an institutional trailblazer in terms of its role in carbon markets, and by others as an impediment to the functioning of these markets. Under the first phase of the KP, China and the five other major emitters (as referenced on a national aggregate, not a per-capita basis), the BASIC countries, were not required to accept mandatory CO_2 emissions caps. Interestingly, Babiker *et al.* (2000) argued that if OECD countries had directly compensated developing countries for losses, rather than engage in the KP, the required annual financial transfer would be on the order of $25 billion (1995 US$) in 2010.

An examination of the KP reveals that it contains no references to "poverty" at all, which indicates that poverty reduction considerations were not seen as relevant to the mandate and focus of the KP. While this may come as no surprise, given that the KP's main objective is to commit Annex B Parties or Annex I Parties to the UNFCCC, what is interesting to note is that the KP contains a sum total of seven references to "energy" – just one additional reference more than the UNFCCC. Just like the UNFCCC, the KP contains no references to "energy access" or "access to energy". Of the seven references to "energy" in total, the first three are all contained in the first paragraph of Article 2, which lists a range of actions that developed countries should take in order to promote sustainable development related to the achievement of their quantified emission limitation and reduction commitments (QELRCs).

Energy is first mentioned in Article 2.1.a (i), which calls for "enhancement of energy efficiency in relevant sectors of the national economy". The second reference contained in Article 2.1.a (iv) calls for "research on, and promotion, development and increased use of, new and renewable forms of energy, of carbon dioxide sequestration technologies and of advanced and innovative environmentally sound technologies". The third reference contained in Article 2.1.a (viii) calls for "limitation and/or reduction of methane emissions through recovery and use in waste management, as well as in the production, transport and distribution of energy (Climate Change Secretariat, 1998, pp. 4–5).

The next reference to energy in the KP is the only one that includes some linkage between climate change mitigation and energy and is contained in Article 10. Article 10 calls on all Parties to formulate and cooperate on a range of activities, but also invokes the principle of "common but differentiated responsibilities" and is explicit in its statement about not introducing any new commitments for Parties no included in Annex I (developing countries). Article 10.b calls for the formulation, implementation, publishing, and updating of national and relevant regional programmes containing climate change mitigation and adaptation measure and, more specifically, Article 10.b.(i). states that such programs would, amongst other things, "concern the energy, transport and industry sectors as well as agriculture, forestry and waste management" (Climate Change Secretariat, 1998, p. 14). The remaining three references include a reference in Article 13 of the KP, but here the reference is with regard to the listing of the International Atomic Energy Agency as an observer along with UN's other specialized agencies. The final two references are both contained in Annex A of the KP, which lists the GHGs covered by the KP and the sector/source categories, which include two references to energy as can be seen in Box 2.1.

Box 2.1   Annex A of the Kyoto Protocol

Greenhouse gases
Carbon dioxide (CO_2)
Methane (CH_4)
Nitrous oxide (N_2O)
Hydrofluorocarbons (HFCs)
Perfluorocarbons (PFCs)
Sulphur hexafluoride (SF6)

Sectors/source categories
Energy
- Fuel combustion
  - Energy industries
  - Manufacturing industries and construction
  - Transport
  - Other sectors
  - Other
- Fugitive emissions from fuels
  - Solid fuels
  - Oil and natural gas
  - Other

Industrial processes
- Mineral products
- Chemical industry
- Metal production
- Other production
- Production of halocarbons and sulphur hexafluoride
- Consumption of halocarbons and sulphur hexafluoride
- Other

Solvent and other product use
Agriculture
- Enteric fermentation
- Manure management
- Rice cultivation
- Agricultural soils
- Prescribed burning of savannas
- Field burning of agricultural residues
- Other

Waste
- Solid waste disposal on land
- Wastewater handling
- Waste incineration
- Other

Source: Climate Change Secretariat (1998, p. 27).

2.3.2   The middle years: Examining the Bali Road Map and the Copenhagen Accord for references to the energy and poverty reduction nexus, and to energy access for the poor

The first COP serving as Meeting of the Parties to the KP – whose official acronym is CMP-1– met in Montreal in 2005 and decided to establish the Ad Hoc Working Group on Annex I Parties' Further Commitments under the Kyoto Protocol (AWG-KP). This Working Group was created in accordance with the KP's Article 3.9, which mandated consideration of Annex I Parties' further commitments at least 7 years before the end of the first commitment period, which was at the end of December 2012. This working group would eventually have a partner working group called the Ad Hoc working Group on Long-Term Cooperative Action (AWG-LCA) as a result of the 2007 agreement reached in the 13th session of the COP to the UNFCC. For those well versed in the language of the climate change negotiations, these acronyms may be easier to comprehend than they are for those outside of the process, but the simplest way to understand the rationale for these two groups is that one group was focused on discussing and negotiating the manner and extent of developed countries' commitments under the second phase of the KP (i.e., the post-2012 version), while the other group was focused on discussing and negotiating the longer-term possibilities and modalities of broader participation by all countries, including developing countries.

Bali Road Map   The 13th session of the COP (COP-13) and the third session of the CMP (CMP-3) met in Bali, Indonesia (December 2007) and agreed to the adoption of the historic Bali Road Map. The Bali Road Map included the Bali Action Plan (BAP) and established key elements for the climate change negotiations process leading up to 2012 and beyond (see Box 2.2). The BAP comprised five sections, with the first being a shared vision that refers to a long-term vision for action on climate change, including the much-anticipated long-term goal for emission reductions. The remaining portion of the BAP was centered around four main building blocks – mitigation, adaptation, technology and financing – and listed a host of issues that needed to be negotiated on in relation to each of the building blocks (UNDP, 2008). Most significantly, the BAP clearly referenced the agreement by all Parties that the negotiations should address a shared vision for long-term cooperative action, including "a long-term global goal for emission reductions", and should "launch a comprehensive process" that would enable the full, effective, and sustained implementation of the Convention, "in order to reach an agreed outcome and adopt a decision at its fifteenth session" (UNFCCC, 2008, p. 3).

Box 2.2   Decisions adopted by the 13th Conference of the Parties (COP-13) in Bali, 2007

1/CP.13	Bali Action Plan
2/CP.13	Reducing emissions from deforestation in developing countries: approaches to stimulate action
3/CP.13	Development and transfer of technologies under the Subsidiary Body for Scientific and Technological Advice
4/CP.13	Development and transfer of technologies under the Subsidiary Body for Implementation
5/CP.13	Fourth Assessment Report of the Intergovernmental Panel on Climate Change
6/CP.13	Fourth review of the financial mechanism
7/CP.13	Additional guidance to the Global Environment Facility
8/CP.13	Extension of the mandate of the Least Developed Countries Expert Group
9/CP.13	Amended New Delhi work programme on Article 6 of the Convention
10/CP.13	Compilation and synthesis of fourth national communications
11/CP.13	Reporting on global observing systems for climate
12/CP.13	Budget performance and the functions and operations of the secretariat
13/CP.13	Programme budget for the biennium 2008–2009
14/CP.13	Date and venue of the 14th and 15th sessions of the Conference of the Parties and the calendar of meetings of Convention bodies

Source: UNFCCC (2008).

COP-13, which adopted the BAP, also decided that the negotiation process for longer-term cooperative action would be conducted within the aegis of the AWG-LCA with a mandate to focus on mitigation, adaptation, finance, technology, and a shared vision for long-term cooperative action. This was the first time that the negotiations had created a specific forum for the discussion of developing countries' roles in addressing future climate change concerns, and so it worth underscoring. Negotiations on Annex I Parties' further commitments continued under the AWG-KP. The noteworthy aspect of the BAP was the institutionalization of a parallel two-track approach for climate negotiations, which was supposed to conclude in at the Copenhagen COP in 2009.

The two-track negotiations process established by the BAP constituted:

- Negotiations related to the UNFCCC under the aegis of the AWG-LCA – included issues related to the four building blocks: adaptation, mitigation, technology transferred and deployment and financing; reducing emissions from deforestation and forest degradation (REDD); nationally appropriate mitigation actions by developing countries; and measurable, reportable, and verifiable nationally appropriate mitigation commitments or actions by all developed countries.
- Negotiations under the Kyoto Protocol under the aegis of the AWG-KP – included developed country emission targets, means to achieve targets, role and use of existing KP mechanisms, market mechanisms, national policies, role of land use, land use change and forestry (LULUCF).

What is interesting to point out is that there are only two references to "poverty eradication" and "poverty reduction", respectively, in the globally agreed decisions emanating from the COP-13 in Bali. The first reference to poverty eradication is contained in Decision 1 (BAP), which reaffirms that economic and social development and poverty eradication are global priorities (UNFCC, 2008, p. 3). The only other reference to poverty reduction is contained in a sub-bullet of a bullet item, i.e., Annex II. 2. 3.d (iii), which does not explicitly focus on climate change mitigation, but instead deals with the "Terms of References of the Expert Group on Technology Transfer". This Expert Group on Technology Transfer would become a highly contested one between developing and developed countries in future COP negotiations, given its potential role in bringing to the fore actions and decision related to technology transfer, which have always been championed by developing countries as far as the 1990s. As per the parlance of climate change negotiations, Annex II calls for the establishment of a proposed Expert group, which would then go on to "propose a two-year rolling programme of work, for endorsement by the twenty-eighth session of the subsidiary bodies following consideration by a joint contact group of the subsidiary bodies", which, amongst other issues, would consider for the medium-term perspective (2008–2012) a set of actions for enhancing the implementation of the technology transfer framework, including the "better integration of national strategies for sustainable development and poverty reduction, based on the United Nations Millennium Development Goals" (UNFCCC, 2008, p. 23). The paragraph referencing poverty reduction has numerous complex suppositional phrases, and may be seen as an example of the complicated negotiated language that has become de rigueur for the UNFCCC outcome documents. The fact that poverty eradication finds a singular mention in the official text of the decision related to the Bali Road Map but is not expressly linked to policy and programmatic inputs or guidance related to climate change mitigation or adaptation, and that poverty reduction is only referenced once within the limited context of a sub-item of a bullet in Annex II can be seen as evidence of the lack of concrete linkages between joint poverty reduction and climate change actions with the text of agreed globally negotiated outcome in Bali.

There are no references to "energy access" or "access to energy" in the Bali decisions. There are nine references to the term "energy" in the agreed text of the Bali decisions, which are also worth considering. Interestingly, the first three references to energy are contained in Decision 3/CP.13, Development and transfer of technologies under the Subsidiary Body for Scientific and Technological Advice, but once again, these references are not related to any specific policy or programmatic guidance on energy linkages with climate change. Instead, these three references relate non-UN partnerships, more specifically to the "progress made by Parties included in Annex II to the Convention in establishing innovative financing partnerships such as the Global Energy Efficiency and Renewable Energy Fund and the European Union Energy Initiative" (UNFCCC, 2008, p. 12).

The next four references to "energy" are all contained in one sub-paragraph of Annex I, which lists recommendations for enhancing the implementation of a technology transfer framework that would facilitate transfer of technologies from Annex I (developed countries) to developing countries. More specifi- cally, the references to energy are contained in the segment dealing with enabling environments for technology transfer, but again, interestingly, the explicit reference is not to any specific policy action or guidance related to energy *per se*, but instead to the possibility of close cooperation with public and/or private partnerships and initiatives established by the World Summit on Sustainable Development and other processes such as "Renewable Energy and Energy Efficiency Partnership, Johannesburg Renewable Energy Coalition, Carbon Sequestration Leadership Forum, and CTI and other International Energy Agency implementing agreements" (UNFCCC, 2008, p. 18). The role of these innovative energy partnerships that have emerged outside of the formal UN-led climate change negotiations process will be discussed further in Chapter 4 in terms of offering new opportunities for enhancing synergies between climate change and energy.

The final two references to "energy" in the Bali decisions again do not pro- vide any specific policy or programmatic guidance on energy-related climate change mitigation or action, but instead are related to the listing of the International Energy Agency's (IEA) role in terms of the EGTT exploring cooperation with other intergovernmental processes to look for possible syn- ergies on technology transfer (UNFCCC, 2008, p. 20). The last reference to energy is also related to the IEA's role, but this time in terms of its membership in the Expert Group on Technology Transfer (EGTT). What is interesting to highlight is that the EGTT emerged as one of key areas of discussion and debate in the ensuing COPs, particularly given the centrality of technology transfer as a key mechanism for developing and developed countries to interact with each other as they engage with climate change mitigation and adaptation.

The Copenhagen Accord   The year 2009 was billed as the year of concrete action on global climate change, and the tag-line associated with the intergovernmental climate change negotiations was that negotiators would "seal the deal" at the 15th COP in Copenhagen. The stakes were high, as COP-15 was charged with the responsibility of adopting a binding global agreement that would govern the second phase of the KP. But the 2-year negotiating process established via the BAP in 2007 did not result in an agreed outcome focused on the global goal for emissions reductions in COP-15 in Copenhagen.

Negotiations under the AWG-KP and AWG-LCA were stalemated over the issue of developing countries wanting to see Annex I Parties (developed countries) commit to clear emissions reduction targets and Annex I Parties wanting to see clear signals from US and BASIC countries in terms of what they would be willing to take on in terms of climate mitigation. The deadline for concluding the two-track negotiations was set for COP-15 in Copenhagen in 2009, but was subsequently extended as a consequence of the divisive and fragmented negotiations at the Copenhagen COP. The two-track negotiations process established by the BAP ended up being extended until the 16th COP in Cancun. The AWG-LCA text has been referenced by the International Institute for Sustainable Development (IISD) as being one of the more complex documents in the history of the UNFCCC, comprising close to 200 pages, with thousands of brackets indicating areas of disagreement that would need further discussion and negotiations (IISD, 2009). A search of the AWG-KP and AWG-LCA documentation and reports leading into the penultimate session (October 2009) just prior to the Copenhagen COP revealed no references related to energy technologies or sustainable energy in either sets of text (UNDP/Cherian, 2009, pp. 15–16).

In the end, after a 2-year process of parallel tracked intense negotiations among the Parties to the UNFCCC and its KP, the final outcome introduced towards the very tail end of the long-drawn-out 2009 Copenhagen negotiations was initially issued as a draft decision by the President of the COP. The Copenhagen Accord, as it was known, ended up creating an intergovernmental negotiating furor, with the result that, for the first time, a COP did not end up formally adopting or agreeing to the Copenhagen Accord, but just "taking note" of it in Decision 2/CP.15. According to the analysis provided by the *Earth Negotiations Bulletin* (ENB), which covered the entire COP proceedings, the Copenhagen COP could be characterized by many dramatic events and long procedural hurdles, including especially the proposal by the Danish President to put forward two texts of his own relating to the AWGs, which angered developing countries and led to concerns about transparency and inclusion (IISD/ENB, 2009).

The Copenhagen Accord merits careful consideration in terms of both its process and final outcomes. Interestingly, the Copenhagen Accord, which ignited a legal debate as to its standing, is evidence of the deal-making role of a new negotiating power broker grouping of countries – the BASIC countries – within the Group of 77 and China. In fact, the BASIC group of countries were not only credited with brokering the final Copenhagen Accord but also met after the Copenhagen COP to coordinate their positions on the Copenhagen Accord (Murray, 2010). The most definitive aspect of the Copenhagen Accord was that the achievement of the ultimate objective of the Convention (Article 2), was linked for the first time with a specific temperature threshold. The first paragraph of the Copenhagen Accord recognized climate change to be one of the greatest challenges of our times that would need to be combated on the basis of the principle of common but differentiated responsibilities, and noted:

> To achieve the ultimate objective of the Convention to stabilize greenhouse gas concentration in the atmosphere at a level that would prevent dangerous anthropogenic interference with the climate system, we shall, recognizing the scientific view that the increase in global temperature should be below 2 degrees Celsius, on the basis of equity and in the context of sustainable development, enhance our long-term cooperative action to combat climate change (UNFCCC, 2010, p. 5).

But Copenhagen did not result in a consensus-based agreement focused on emission targets and adaptation action. The concluding segment of the Copenhagen COP comprised a series of lengthy, closed-door, and tense negotiations characterized by the IISD Reporting Services as "acrimonious", with questions raised about the transparency of the negotiations process (IISD Reporting Services, 2009). Finally, after meeting all night and after an arduous 13-hour plenary session, Parties agreed to adopt a COP decision whereby the COP agreed to "take note" of the Copenhagen Accord, which was attached to the decision as an unofficial document but was not formally agreed on or adopted by the COP. The Copenhagen COP revealed an important faultline separating the global rhetoric of pushing for a global climate change deal versus the actual status of intergovernmental climate change negotiations, because in the end the much anticipated Copenhagen Accord did not include any concrete emissions targets. To date, the UNFCCC website in its home page lists under the tabular title of "key steps", The Convention, Kyoto Protocol, Bali Road Map, Cancun Agreements, Durban Outcomes, Doha Climate Gateway and Warsaw Outcomes, but there is no tabular reference or mention to the 2009 Copenhagen Accord (UNFCCC website, 2015).

The issue of "poverty eradication" is mentioned just once in the entire set of decisions that were agreed to at the Copenhagen COP-15:

> We agree that deep cuts in global emissions are required according to science.... We should cooperate in achieving the peaking of global and national emissions as soon as possible, recognizing that the time frame for peaking will be longer in developing countries and bearing in mind that social and economic development

and poverty eradication are the first and are the first and overriding priorities of developing countries and that a low-emission development strategy is indispensable to sustainable development (UNFCCC, 2010, pp. 5–6).

Although the reference segues from the prioritization of poverty eradication to the idea that low emission development is part and parcel of sustainable development, there is no actual policy guidance or concrete linkages between energy and climate change that could facilitate any such low-emission development pathway, nor is there any specific mention of sustainable energy access for the poor initiatives that could simultaneously address socio-economic development and poverty reduction concerns.

The Copenhagen COP outcomes contain no references to "energy access" or "access to energy". More surprisingly, the term "energy" is only mentioned once in the entire set of decisions comprising the outcome document. The sole reference to energy does not include any stated or clear policy guidance or input on energy sector mitigation, but instead is contained in a sub-item of an Annex entitled, "Updated training programme for greenhouse gas inventory review experts for the technical review of greenhouse gas inventories of Parties included in Annex I to the Convention". Here, the specific reference to energy is in association with a basic course outline for the inventory of GHGs for Annex I countries, which list IPCC guidelines on various sectors including energy (UNFCCC, 2010, p. 27).

2.3.3 The recent years: Examining the Cancun Agreements, Durban Outcomes, Doha Climate Gateway, the Warsaw Outcomes and the LCCA for references to the energy and poverty reduction nexus, and to energy access for the poor

Cancun Agreements The 16th session of the COP (COP-16) met in Cancun, Mexico, at end of 2010. According to the UNFCCC, the Cancun Agreements included the most comprehensive package ever agreed by governments to help developing nations deal with climate change- encompassing finance, technology, and capacity-building support. Decision 1/CP.16 recognized the need for deep cuts in global emissions in order to limit the global average temperature rise to 2°C above pre-industrial levels. Parties agreed to consider strengthening the global long-term goal during a review by 2015, including in relation to a proposed 1.5°C target. This meant that the Cancun Agreements were the first official UN agreement related to the 2°C temperature threshold, which was referenced in the Copenhagen Accord. The Cancun Agreements also established several new institutions and processes, including the Cancun Adaptation Framework, the Adaptation Committee, and the Technology Mechanism, comprising the Technology Executive Committee (TEC) and the Climate Technology Centre and Network (CTCN) and the Green Climate Fund (GCF), which was designated as an operating entity of the Convention's financial mechanism governed by a 24-member board.

The ENB analysis of COP-16 noted that the expectations for Cancun were modest, especially in the light of the haunting specter of the Copenhagen negotiations which were characterized by "mistrust, confusion and parallel discussions by experts and Heads of State and Government". According to ENB, Mexico managed a "disciplined and extensive campaign" which included their commitment to a "transparent and inclusive" process during the negotiations, which "many saw restoring faith in the process and laying to rest the ghosts of Copenhagen as the most important achievement" (ENB, 2010). Stavins (2010) points to the most notable outcomes of Cancun Agreements being:

1. Provision of emission mitigation targets and actions, including, most significantly, pledges by the largest aggregate emitters to reduce emissions by 2020, and the first official agreement to keep temperature increases below a global average of 2°C.
2. Inclusion of mechanisms for monitoring and verification, including independent expert panel analyses of developing country mitigation actions.
3. The establishment of the GCF, with the World Bank designated as an interim trustee

Along the same lines, Dodds *et al.* (2014) have stated that the three accomplishments of the Cancun COP were an inclusive process of outreach and consultations, the universal recognition and commitment of a climate temperature threshold of 2°C, and the establishment of a GCF.

What was different about the Cancun COP is that there was a broader approach to emission reductions for the first time in the history of the climate change negotiations. Article 4.7 of the Convention established that mitigation actions taken by developing countries would be supported and the BAP for the first time referenced the idea that developing countries would submit nationally appropriate mitigation actions (NAMAs). But the Cancun Agreement broadened the actual scope of participation in climate change mitigation action. Developed countries agreed to economy-wide emission reduction targets and strengthened reporting frequency and standards. The Cancun Agreement "emphasized the need for deep cuts" in GHG emissions and "early and urgent undertakings to accelerate and enhance the implementation of the Convention by all Parties" (UNFCCC, 2011, p. 7). Developing countries agreed to the submission of NAMAs to be implemented subject to financial and technical support. Although the agreement referenced the issue of equity and the overriding importance of poverty eradication and the socio-economic development space for developing countries, paragraph 48 on the NAMAs can be seen as proverbial game-changer terms of broadening the scope of mitigation action. The Cancun Agreements articulated for the first time the idea that developing countries through their NAMAs could be proactive in emissions reductions. In other words, there was agreement that "developing country Parties will take nationally appropriate mitigation actions in the context of sustainable development, supported and enabled by technology, financing and capacity-building, aimed at achieving a deviation in emissions relative to 'business as usual' emissions in 2020" (UNFCCC, 2011, pp. 9–10).

All the agreed decisions emanating from COP-16 are listed in Boxes 2.3 and 2.4 and represent the Cancun Agreements. Box 2.4 lists all the COP-16 decisions, including the most significant outcome – Decision 1/CP.16 – entitled "Outcome of the work of the Ad Hoc Working Group on Long-term Cooperative Action under the Convention" (UNFCCC, Addenum 1, 2011). All the remaining 11 decisions agreed to the Cancun COP are contained in a separate addendum outcome document and are also listed in the box.

An examination of the 12 decisions from COP-16 indicates that there are no references to "energy". That is to say, there is absolutely no mention made of "energy" in any of these Cancun COP decisions, which is quite surprising given the important role of energy in current and future climate change-related GHG emissions, and the overall focus on longer-term cooperative climate action, which is anticipated to include a broad range of countries. Given that energy is not mentioned once in the Cancun COP outcomes, it is not surprising that references to "access to energy" and "energy access" are also missing from these outcomes.

Box 2.3   Cancun Agreements – key features related to developing countries' nationally appropriate mitigation actions (NAMAs)

- Developing countries will provide information on the NAMAs for which they are seeking support, whereas industrialized countries will provide information on available support for these actions.
- Supported actions will be measured, reported and verified internationally, whereas for domestically supported actions this will be done at the national level. The intention is that the countries which provide the support, and the countries which receive the support, are both satisfied that adequate resources are going to the right place for the right reasons and are having the best impact.
- It was also agreed that developing countries will also increase reporting of progress towards their mitigation objectives, although in a differentiated way to that of industrialized countries. A process of international consultation and analysis of these biennial reports will be established.
- Developing countries are encouraged under the agreement to draw up low-carbon development strategies or plans.

Source: UNFCCC. Available at http://cancun.unfccc.int/mitigation/decisions-addressing-developing-country-mitigation-plans/#c178.

Box 2.4   *Decisions taken at Cancun Climate Change 16th Conference of the Parties (COP-16)*

1/CP.16   The Cancun Agreements: Outcome of the work of the Ad Hoc Working Group on Long-term Cooperative Action under the Convention

2/CP.16   Fourth review of the financial mechanism

3/CP.16   Additional guidance to the Global Environment Facility

4/CP.16   Assessment of the Special Climate Change Fund

5/CP.16   Further guidance for the operation of the Least Developed Countries Fund

6/CP.16   Extension of the mandate of the Least Developed Countries Expert Group

7/CP.16   Progress in, and ways to enhance, the implementation of the amended New Delhi work programme

8/CP.16   Continuation of activities implemented jointly under the pilot phase

9/CP.16   National communications from Parties included in Annex I to the Convention

10/CP.16   Capacity-building under the Convention for developing countries

11/CP.16   Administrative, financial and institutional matters

12/CP.16   Dates and venues of future sessions

Source: UNFCCC (2011a,b).

The Cancun Agreement (Decision 1/CP.16) contains five references to "poverty eradication". The first reference is contained in the opening section which affirms the needs of developing country Parties for "the achievement of sustained economic growth and the eradication of poverty, so as to be able to deal with climate change" (UNFCCC, 2011, p. 2) The next reference is contained in the "shared vision" section of the document, and reiterates the language of the Copenhagen Accord calling on parties to cooperate in "peaking of global and national" GHG emissions as soon as possible, recognizing that the time frame for peaking will be "longer in developing countries" bearing in mind that socio-economic development and "poverty eradication are the first and overriding priorities of developing countries" (UNFCCC, 2011, p. 3). So here the idea is that poverty eradication is necessarily to be prioritized over climate mitigation actions by developing countries. The third reference is contained in Section III, focused on enhanced action on mitigation by developing countries – namely, the NAMAs – and here again the issue of socio-economic development and poverty eradication being the first and overriding priorities is referenced along with the idea that "the share of global emissions originating in developing countries will grow to meet their social and development needs (UNFCCC, 2011, p. 9). The fourth reference is included in the section dealing with economic and social consequences of response measures, where it is affirmed that responses to climate change should be coordinated with socio-economic development amongst other issues, taking fully into account the legitimate priority needs of developing country Parties for the achievement of sustained economic growth and the eradication of poverty (UNFCCC, 2011, p. 15). The final reference to "reducing poverty" is contained in an appendix relating to the policy guidance on reducing emissions from deforestation (UNFCCC, 2011, p. 26).

Box 2.5 contains all decisions emanating from the COP serving as the sixth meeting of the Parties to the Kyoto Protocol (CMP-6). There are no references to "poverty" and no linkages whatsoever to poverty reduction or eradication in the remaining Decisions 2–12. In addition to the decisions

---

Box 2.5   Decisions adopted by the 16th Conference of the Parties (COP-16) in Cancun, serving as the meeting of the Parties to the Kyoto Protocol at its sixth session (CMP-6)

1/CMP.6   The Cancun Agreements: Outcome of the work of the Ad Hoc Working Group on Further Commitments for Annex I Parties under the Kyoto Protocol at its fifteenth session

2/CMP.6   The Cancun Agreements: Land use, land-use change and forestry

3/CMP.6   Further guidance relating to the clean development mechanism

4/CMP.6   Guidance on the implementation of Article 6 of the Kyoto Protocol

5/CMP.6   Report of the Adaptation Fund Board

6/CMP.6   Review of the Adaptation Fund

7/CMP.6   Carbon dioxide capture and storage in geological formations as clean development mechanism project activities

8/CMP.6   Proposal from Kazakhstan to amend Annex B to the Kyoto Protocol

9/CMP.6   Methodology for the collection of international transaction log fees in the biennium 2012–2013

10/CMP.6   Supplementary information incorporated in national communications submitted in accordance with Article 7, paragraph 2, of the Kyoto Protocol

11/CMP.6   Capacity-building under the Kyoto Protocol for developing countries

12/CMP.6   Administrative, financial and institutional matters

13/CMP.6   Compliance Committee

Source: UNFCCC/CMP (2011a,b).

emanating from the COP, there were also a series of decisions that were agreed to by CMP-6 at Cancun which included the "Outcome of the work of the Ad Hoc Working Group on Further Commitments for Annex I Parties under the Kyoto Protocol at its fifteenth session" (Decision 1/CMP.6). Box 2.5 itemizes the complete list of all the CMP decisions and these are directly excerpted from two CMP-6 outcome documents, namely Addendum 1 and Addenum 2. There are no references to "energy" or "poverty" in any of the CMP-6 decisions listed in the box (UNFCCC/CMP, 2011a,b).

Durban Outcomes   The 17th session of COP was held in December 2011 in Durban, South Africa, and has been recognized as a watershed in the long trajectory of global climate change negotiations, because in Durban, all governments agreed to commit themselves to a time-bound process to arrive at a global agreement – that is, another 4 years of negotiations to reach a universal legally binding agreement and an explicit time limit of achieving a comprehensive climate deal by December 2015. The Durban COP also made a significant contribution to the two-track process of climate change negotiations initiated as a consequence of the 2007 BAP, in that the Durban decision established a new Ad Hoc Working Group on the Durban Platform for Enhanced Action (ADP) and provides that the existing AWG-LCA would conclude by the end of 2012. According to the UNFCCC, the key outcome was the Durban Road Map for Implementation which had four main areas of focus:

- Second commitment period of the Kyoto Protocol until 2017.
- Launch of new platform of negotiations under the UNFCCC to deliver a new and universal GHG reduction protocol, legal instrument or other outcome with legal force by 2015 for the period beyond 2020.
- Conclusion of existing broad-based streams of negotiation by 2012 (including work to make national emissions reductions more transparent, launch of global support network on technology etc.).
- Global Review scoping and review of emerging climate challenge based on new data to ensure adequacy of the 2°C rise or a requirement of a lower threshold of 1.5°C.

For some, what set the Durban COP apart was its adoption of a time frame for a future climate change agreement billed for 2015, which would bind both developed countries and large emitter developing countries (Toyne, 2011), while others have noted that the language contained could be interpreted as vague but that even these relatively indefinite words required negotiating finesse in terms of being agreeable to those who favor a binding approach versus those who favored a non-binding one (Hultman, 2012). Others have argued that the Durban Platform's agreement to move out of the realm of voluntary action and initiate a legal time-bound framework that could cover all countries is a significant departure, and universal participation along with legally binding mitigation targets could prove to make Durban a landmark conference (Carpenter, 2012). Meanwhile in his analyses of the Durban outcomes, Bodansky (2012) points out that the "Durban Platform is significant not only for what it says, but for what it does not say", citing, for instance, the fact that the Durban Platform makes no reference to the two-track process of negotiations, does not repeat the Convention's language that developed countries should "take the lead" in combating climate change, and contains no reference to developing, developed, Annex I or non-Annex I Parties categories that have dominated the climate change regime thus far (p. 3) But the Durban Platform does not really provide any guidance on the substantive content of negotiations.

An examination of all the decisions agreed at the Durban Conference for references to energy and, more specifically, energy access for the poor indicates that there is no concrete mention of the issue of energy access for the poor within these agreed outputs. All the agreed decisions emanating from Durban COP-17 and all decisions emanating from the COP serving as the seventh meeting of the Parties to the Kyoto Protocol (CMP-7) are listed in Boxes 2.6 and 2.7 and represent the Durban Outcomes. Box 2.6 lists all of the 19 decisions agreed at COP-17, including the most significant outcome – Decision 1/CP.17 – entitled "Establishment of an Ad Hoc Working Group on the Durban Platform for Enhanced Action (UNFCCC, 2012a). There are no references made to "access to energy" or "energy access" in all 19 of the COP-17 decisions.

Decisions 1–19 of COP-17 are contained in two separate outcome documents (Decisions 1–5 in Addendum 1 and Decisions 6–19 in Addendum 2). An examination of Decisions 1–5 of COP-17 reveals three references to "poverty" that are all contained in Decision 2, are in sections analogous to the ones in Cancun Agreements, and also borrow from the previously agreed to text used within the UNFCCC context. That is to say, the first reference, is contained in the section dealing with NAMAs by developing country Parties and the reference makes note of the fact that socio-economic development and "poverty eradication are the first and overriding priorities of developing country Parties" and also mentions that "the share of global emissions in developing countries would grow to meet social and development needs" (UNFCCC, 2012a, p. 9). The second reference is contained in the section dealing with policy approaches to do with reducing emissions from deforestation and forest degradation but is included as a preambular idea that policy approaches related to mitigation in the forest sector "can promote poverty alleviation and biodiversity benefits" (UNFCCC, 2012a, p. 15). The final reference to poverty

Box 2.6   Decisions adopted by the 17th Conference of the Parties (COP-17) in Durban

1/CP.17	Establishment of an Ad Hoc Working Group on the Durban Platform for Enhanced Action
2/CP.17	Outcome of the work of the Ad Hoc Working Group on Long-term Cooperative Action under the Convention
3/CP.17	Launching the Green Climate Fund
4/CP.17	Technology Executive Committee – modalities and procedures
5/CP.17	National adaptation plans
6/CP.17	Nairobi work programme on impacts, vulnerability and adaptation to climate change
7/CP.17	Work programme on loss and damage
8/CP.17	Forum and work programme on the impact of the implementation of response measures
9/CP.17	Least Developed Countries Fund: support for the implementation of elements of the least developed countries work programme other than national adaptation programmes of action
10/CP.17	Amendment to Annex I to the Convention
11/CP.17	Report of the Global Environment Facility to the Conference of the Parties and additional guidance to the Global Environment Facility
12/CP.17	Guidance on systems for providing information on how safeguards are addressed and respected and modalities relating to forest reference emission levels and forest reference levels as referred to in decision 1/CP.16
13/CP.17	Capacity-building under the Convention
14/CP.17	Work of the Consultative Group of Experts on National Communications from Parties not included in Annex I to the Convention
15/CP.17	Revision of the UNFCCC reporting guidelines on annual inventories for Parties included in Annex I to the Convention
16/CP.17	Research dialogue on developments in research activities relevant to the needs of the Convention
17/CP.17	Administrative, financial and institutional matters
18/CP.17	Programme budget for the biennium 2012–2013
19/CP.17	Dates and venues of future sessions

Source: UNFCCC (2012a,b).

Box 2.7   *Decisions adopted by the 17th Conference of the Parties (COP-17) in Durban, serving as the meeting of the Parties to the Kyoto Protocol at its seventh session (CMP-7)*

1/CMP.7   Outcome of the work of the Ad Hoc Working Group on Further Commitments for Annex I Parties under the Kyoto Protocol at its sixteenth session

2/CMP.7   Land use, land-use change and forestry

3/CMP.7   Emissions trading and the project-based mechanisms

4/CMP.7   Greenhouse gases, sectors and source categories, common metrics to calculate the carbon dioxide equivalence of anthropogenic emissions by sources and removals by sinks, and other methodological issues

5/CMP.7   Consideration of information on potential environmental, economic and social consequences, including spillover effects, of tools, policies, measures and methodologies available to Annex I Parties

6/CMP.7   Report of the Adaptation Fund Board

7/CMP.7   Review of the Adaptation Fund

8/CMP.7   Further guidance relating to the clean development mechanism

9/CMP.7   Materiality standard under the clean development mechanism

10/CMP.7   Modalities and procedures for carbon dioxide capture and storage in geological formations as clean development mechanism project activities

11/CMP.7   Guidance on the implementation of Article 6 of the Kyoto Protocol

12/CMP.7   Compliance Committee

13/CMP.7   Proposal from Kazakhstan to amend Annex B to the Kyoto Protocol

14/CMP.7   Appeal by Croatia against a final decision of the enforcement branch of the Compliance Committee in relation to the implementation of decision 7/CP

15/CMP.7   Capacity-building under the Kyoto Protocol

16/CMP.7   Administrative, financial and institutional matters

17/CMP.7   Programme budget for the biennium 2012–2013

Source: UNFCCC/CMP (2012a,b).

is contained in paragraph 87 of the section dealing with "economic and social consequences of response measures", which states that "social and economic development and poverty eradication are the overriding priorities of developing countries" (UNFCCC, 2012a, p. 18).

There is only one reference to "energy" in Decisions 1–5 of the Durban COP-17. The one reference does not provide any specific policy guidance in terms of actual mitigation options, nor is it mentioned in connection with poverty reduction or energy access issues. The sole reference is contained in a sub-paragraph of a sub-section of an Annex, which discusses progress in achievement of quantified economy-wide emission reduction targets and relevant information to be provided by Annex I Parties (developed countries), which in this case states "To the extent appropriate, Parties shall organize the reporting of mitigation actions by sector (energy, industrial processes and product use, agriculture, LULUCF, waste and other sectors) and by gas (carbon dioxide, methane, nitrous oxide, hydrofluorocarbons, perfluorocarbons and sulphur hexafluoride)" (UNFCCC, 2012a, p. 32).

Meanwhile an examination of Decisions 6–19 of COP-17 indicates that there are no references to "poverty" in any of these decisions. Decisions 6–19 contain six references to "energy", which are all found in Decision 15 dealing with the revision of the UNFCCC reporting guidelines on annual inventories for Parties included in Annex I to the Convention. The first two references are contained in paragraph 35, but in fact the first reference deals with the term "non-energy": "Annex I Parties should clearly indicate how feedstocks and non-energy use of fuels have been accounted for in the inventory, under the energy or industrial processes sector, in accordance with the 2006 IPCC Guidelines" (UNFCCC, 2012b, p. 33). The third reference is found in paragraph 48 dealing with the national inventory report, but once again it deals with "information on how and where feedstocks and non-energy use of fuels have been reported in the inventory" (UNFCCC, 2012b, p. 36). The next set of references deal with the topic of energy pertaining to the outline and general structure of the national inventory, but one of these again deals with "feedstocks and the non-energy use of fuels" (UNFCCC, 2012b, p. 39) and the final reference deals with "Annex 4: The national energy balance for the most recent inventory year" (UNFCCC, 2012b, p. 43).

Box 2.7 lists all 17 of the CMP-7 decisions taken at the Durban COP. There is no mention of "poverty" in any of the 17 decisions of CMP-7. There are also no references to "access to energy" or "energy access for the poor" in any of the CMP-7 decisions (UNFCCC/CMP, 2012a,b). There are four references to "energy" and these references are clustered in two decisions of CMP-7. There is one reference to "energy" in Decision 2 dealing with land use, land-use change, and forestry. What is interesting to note is that this sole reference to energy is not in connection with fossil fuel or energy-related GHGs, but is contained in paragraph 32 of subsection E of the Annex focused on the definitions, modalities, rules, and guidelines relating to land use and forestry: "Carbon dioxide emissions from wood harvested for energy purposes shall be accounted for on the basis of instantaneous oxidation" (UNFCCC/CMP, 2012a, p. 17).

The remaining three references to energy are contained in Decision 8, which provides guidance to the Clean Development Mechanism. The first two of these references are found in paragraph 17, which:

> Encourages the Executive Board to extend the simplified modalities for the demonstration of additionality to a wider scope of project activities, inter alia energy-efficiency project activities and renewable energy based electrification in areas without grid connection, and to develop simplified baseline methodologies for such project activities (UNFCCC/CMP, 2012b, p. 7).

The final reference on energy has to do with capacity-building and discusses support for "microscale renewable energy technologies that are automatically defined as additional" (UNFCCC/CMP, 2012b, p. 9).

After 20 years of negotiations, the adoption of the Durban Platform for Enhanced Action, which aims to ensure "the highest possible mitigation efforts by all Parties", constituted a unique opportunity to raise the level of ambition to address climate change. From the perspective of the Durban outcomes, non-Annex I (developing country). Parties were "invited" to align themselves with the Cancun Agreement by reporting their NAMAs. Analyses of these NAMAs are important because these are anticipated to be the main mechanism for mitigation action in developing countries in any future post-2012 climate change agreement. While NAMAs will in all likelihood be a key modality/mechanism for understanding and addressing developing country Parties' emissions reductions, the intergovernmental negotiations have not provided clear and concise guidelines as to all the informational aspects related to the NAMAs, nor have the key elements of the measurement, reporting, and verification (MRV) process and mechanisms been outlined and agreed to, and clearly this is an area of much debate and future work. Ideas related to a template for NAMAs have been provided by Lütken *et al.* (2011) and the UNFCCC (2012) and the International Renewable Energy Agency (IRENA, 2012) have suggested elements for a process to prepare NAMAs. Sharma and Desgain (2013) point out that, since the scope of the NAMAs can currently range from broad (strategy, policy/regulation) to narrow (project), efforts related to designing NAMAs hinge on a variety of factors, from the planning process to the data and information needed for the design, and the data and information required for monitoring and evaluating the implementation.

The most up-to-date analysis of NAMAs by developing countries has been compiled by the UNFCCC Secretariat in a document entitled "Compilation of information on nationally appropriate mitigation actions to be implemented by developing country Parties" (UNFCCC/SBI, 2013). The 60-page report contains 257 references to "energy", which includes numerous references to improving energy efficiency and renewable energy as well as increasing energy standards and the overall role of the energy sector by a variety of developing countries' NAMAs. However, there are only two references made to the concept of "access to energy". The first is made in connection to the NAMA by Burkina Faso, which highlights the role of access to energy in connection to reducing poverty and meeting the Millennium Development Goals (UNFCCC/SBI, 2013, p. 12), and the second is made in connection to the NAMA by Guinea, which discusses the role of access to energy in rural areas keeping in mind reduction of GHG emissions (UNFCCC/SBI, 2013, p. 49). The compilation report on NAMAs contains five references to "poverty" and includes the African Group's emphasis on the importance of food security, poverty eradication, and enhanced socio-economic development amongst other concerns (UNFCCC/SBI, 2013, p. 4) and four countries – Botswana, Burkina Faso, Georgia and Paupa New Guinea – referencing poverty eradication as a concern. In the absence of clear and comprehensive agreement on guidance related to energy for sustainable development objectives pertaining to NAMAs, the linkages between energy, poverty reduction and climate change are not evident in this compilation report.

Doha Gateway   The 18th session of the COP (COP-18) to the UNFCCC and the eighth session of the CMP (CMP-8) was held from November 26 to December 8, 2012, in Doha, Qatar. Like previous COPs, there was a lengthy and arduous negotiating push towards the final hours of the conference, which was officially gaveled to a close after a 36-hour final set of negotiation sessions amongst 195 countries in attendance. Essentially the carry-over from Durban had created three parallel negotiating tracks – the AWG-LCA, the AWG-KP and the ADP (whose first chair came from India) – but clearly having three negotiating tracks was unwieldy and cumbersome even for seasoned climate change negotiations veterans and particularly challenging and difficult for small country delegations, especially given the propensity for late and all-night climate change negotiations sessions in all the COPs. Like Cancun and Durban, the Doha negotiations may broadly be characterized as incrementally building on resolving the outstanding issues that were being considered by the parallel track negotiations. Finally, the AWG-LCA that was supposed to end at the fractured and tense 2009 Copenhagen COP-15 but ended being extended to COP 17 was terminated at COP18 as previously agreed to in COP-17.

A significant result of the Doha negotiations was that the parallel track process was streamlined as both the AWG-LCA and the AWG- KP finalized their work in Doha. Outstanding issues where agreement was not reached were divided up and allocated to the regular UNFCCC bodies – the Subsidiary Body for Scientific and Technological Advice (SBSTA) and the SBI. The ADP, agreed in Durban in 2011, was officially mandated to take on two workstreams: Workstream 1 was designated to take the steps necessary to negotiate a global climate change agreement that will be adopted by 2015 and enter into force from 2020; and Workstream 2 was designated to agree on how to raise global ambition before 2020 to accelerate the response to climate change. Another key challenge was the issue of climate financing – which has long been a bone of contention between developing and developed countries. Noting that $30 billion in grants and loans promised in 2009 were set to expire in 2012, and that a GCF designed to channel up to $100 billion annually to poor countries has yet to begin operating, the lack of clarity and agreement on the scope and nature of financing was seen as key factor under consideration by developed and developing countries in Doha (Ritter, 2012), but in the end the Doha agreement was suitably vague on this topic, preferring to push the discussions on to Warsaw.

The most immediately notable result emanating from Doha was the agreement on a second phase of the KP. With the first phase of the KP set to expire on December 31, 2012, and after years of negotiations starting from 2005 under the aegis of the AWG-KP, in the final hours of Doha, there was agreement on the extension of the KP into a second commitment phase that is a post-2012 period. The first phase of KP had binding targets for 37 industrialized countries and the EU that were supposed to result in emission reductions of 5% below 1990 levels in 2008–2012, a figure that was not deemed ambitious when it was adopted in 1997 and would end up being critiqued heavily since then, given the non-participation of the US. But the subsequent abandonment of the second phase of the KP by key developed country parties, including Australia, Canada, New Zealand, the Russian Federation, and Japan, the host country of the original KP, points to this second phase becoming a mere shadow of the former phase, with these countries refusing to take on commitments in the second KP commitment period. The Doha amendment to the KP has not yet entered into force and a total of 144 instruments of acceptance are required for its entry into force.

In its summary analysis of the Doha COP, the ENB notes that the current coverage of the KP only covers about 15% of global GHG emissions, and that the average 18% emission reduction by Annex I parties from 1990 levels in 2013–2020 is not nearly enough to put the world on track to avoid the 2°C temperature increase limit (ENB, 2012). The COP was not without drama as tense negotiations ensued over the issue of assigned amount units

(AAUs) of the KP and whether they could be carried over to the second phase. The ENB notes that Russia, Ukraine, and Belarus attempted to block the adoption of the AWG-KP outcome, which included limits on the amount of surplus AAUs that can be used, during the CMP final plenary, but the COP President gaveled its adoption before appearing to notice Russia's raised flag. Russia's objection to the breach of procedure and the COP President's response that the Russian views would be reflected in the final report, according to the ENB, echoed Cancun COP's closing plenary when Bolivia's objections to the adoption of the Cancun Agreement were over-ruled in much the same way, with ENB raising the question as to whether this was becoming trend in the climate negotiations, keeping in mind that a consensus does not mean the right of one Party to the UNFCC to block agreement (ENB, 2012).

In terms of climate change mitigation from the perspective of developing countries, at COP-18 in Doha, Parties agreed to establish a work program that would focus primarily on the information needed to improve under-standing and knowledge about NAMAs (including the underlying method-ologies and assumptions for estimation of mitigation impacts) and the need for NAMAs to get support in terms of preparation and implementation and the role of the Registry in terms of matching NAMAs with global support. But as Sharma and Desgain (2013) point out, financing NAMAs is likely to be a combination of the following four sources: international public finance including the GCF, domestic public finance, private sector investments, and national and international commercial financial institutions. They argue that the share of international or domestic public finance will be high in NAMAs that are aimed at creating conducive policy and regulatory environment for channeling investments to low GHG emissions options or institutional strengthening. In the case of NAMAs that directly support implementation of mitigation options, the share of public finance will be smaller, and the private sector will have a greater role to play. But the fact of the matter remains that developing countries are faced with the challenges of high degrees of variations in terms of NAMA submissions, and with a lack of policy guidance and institutional clarity on how best the matching functions of the Registry will provide guidance and the right mix of financing for individual NAMAs.

Another significant outcome from the Doha COP from a developing country perspective was a new discussion topic promoted by and relating to countries that are globally recognized as being the most vulnerable to the adverse effects of climate change – the issue of "loss and damage". This con-cept emerged for the first time in Doha and refers to the need for compensation to vulnerable countries for their loss and damage as a result of climatic changes. Discussions on this were heated and lengthy and there was no agreement on any institutional mechanism relating to this concept. Pointing out that this was the first time the phrase "loss and damage from climate change" was enshrined in international climate change document, Harvey (2012) noted that the recognition of the need for "damage aid" to poor vulnerable countries "stopped well short of any admission of legal liability or the need to pay compensation of developed countries" and left key questions unaccounted for, such as whether the funds would come from existing donor humanitarian aid budgets and how they would be disbursed. Clearly COP-18 resulted in no less than 26 decisions, as Box 2.8 demonstrates. A review of all the decisions demonstrates that Decisions 1–18 and 20–26 of the Doha COP-18 contain **no** references to "energy" or "poverty" – that is to say, all of these decisions (1–18 and 20–26) contain no mention of "energy access", "access to energy" and "poverty eradication" (UNFCCC, 2013a–c).

There is one decision – Decision 19, contained in Addendum 3 – which is focused on providing a common tabular format for "UNFCCC biennial reporting guidelines for developed country Parties" that contains 22 references to "*energy*". But this decision does not provide guidance on energy access, and references "energy" in the following areas: the category of energy and energy industries (UNFCCC, 2013c, pp. 7–11), as a sector in the list of quantified economy-wide emission reduction targets (UNFCCC, 2013c, p.16), and in the footnotes to tables in which parties are asked to the extent possible to include reporting on energy and other sectors including agriculture and forestry (UNFCCC, 2013c). To be clear, Decision 19 contains no mention of poverty eradication or poverty reduction or energy access for the poor (UNFCCC, 2013c).

A review of all the CMP-8 decisions contained in Box 2.9 reveals that there are **no** references to "poverty" in Decisions 1–13 (UNFCCC/CMP, 2013a–c). There are three references to "energy" in Decisions 1, 2 and 5 of CMP-8. The first mention of energy is contained in Annex 2 of Decision 1, entitled, "Political declarations relating to assigned amount units carried over from the first commitment period of the Kyoto Protocol", but here the reference to energy is to the European Union legislation dealing with "Climate-Energy Package" for the implementation of emission reductions for the period 2013–2020 (UNFCCC/CMP, 2013a, p. 13). The second reference to energy is contained in another annex but this time in Decision 2, and it is focused on information relating to land use and forestry activities of the Kyoto Protocol in annual GHG inventories: … "wood harvested for energy purposes have been accounted on the basis of instantaneous oxidation" (UNFCCC/CMP, 2013a, p. 20). The final reference to energy does not provide guidance or concrete inputs related to energy sector mitigation. Instead, it can be found in yet another Annex in Decision 5 dealing with the

Box 2.8 Decisions adopted by the 18th Conference of the Parties (COP-18) in Doha

1/CP.18	Agreed outcome pursuant to the Bali Action Plan
2/CP.18	Advancing the Durban Platform
3/CP.18	Approaches to address loss and damage associated with climate change impacts in developing countries that are particularly vulnerable to the adverse effects of climate change to enhance adaptive capacity.
4/CP.18	Work programme on long-term finance
5/CP.18	Report of the Standing Committee
6/CP.18	Report of the Green Climate Fund to the Conference of the Parties and guidance to the Green Climate Fund
7/CP.18	Arrangements between the Conference of the Parties and the Green Climate Fund
8/CP.18	Review of the financial mechanism
9/CP.18	Report of the Global Environment Facility to the Conference of the Parties and additional guidance to the Global Environment Facility
10/CP.18	Further guidance to the Least Developed Countries Fund
11/CP.18	Work of the Adaptation Committee
12/CP.18	National adaptation plans
13/CP.18	Report of the Technology Executive Committee
14/CP.18	Arrangements to make the Climate Technology Centre and Network fully operational
15/CP.18	Doha work programme on Article 6 of the Convention
16/CP.18	Prototype of the registry
17/CP.18	Composition, modalities and procedures of the team of technical experts under international consultations and analysis
18/CP.18	Work of the Consultative Group of Experts on National Communications from Parties not included in Annex I to the Convention
19/CP.18	Common tabular format for "UNFCCC biennial reporting guidelines for developed country Parties"
20/CP.18	Status of submission and review of fifth national communications from Parties included in Annex I to the Convention and compilation and synthesis of fifth national communications from Parties included in Annex I to the Convention
21/CP.18	Capacity-building under the Convention for countries with economies in transition
22/CP.18	Activities implemented jointly under the pilot phase
23/CP.18	Promoting gender balance and improving the participation of women in UNFCCC negotiations and in the representation of Parties in bodies established pursuant to the Convention or the Kyoto Protocol
24/CP.18	Economic diversification initiative
25/CP.18	Administrative, financial and institutional matters
26/CP.18	Dates and venues of future sessions

Sources: UNFCCC (2013a–c).

> ## Box 2.9 Decisions adopted by the Conference of the Parties (COP) serving as the meeting of the Parties to the Kyoto Protocol at its eighth session (CMP-8)
>
> 1/CMP.8   Amendment to the Kyoto Protocol pursuant to its Article 3, paragraph 9 (the Doha Amendment)
>
> 2/CMP.8   Implications of the implementation of decisions 2/CMP.7 to 5/CMP.7 on the previous decisions on methodological issues related to the Kyoto Protocol, including those relating to Articles 5, 7 and 8 of the Kyoto Protocol
>
> 3/CMP.8   Report of the Adaptation Fund Board
>
> 4/CMP.8   Initial review of the Adaptation Fund
>
> 5/CMP.8   Guidance relating to the clean development mechanism
>
> 6/CMP.8   Guidance on the implementation of Article 6 of the Kyoto Protocol
>
> 7/CMP.8   Supplementary information incorporated in national communications from Parties included in Annex I to the Convention that are also Parties to the Kyoto Protocol and submitted in accordance with Article 7, paragraph 2, of the Kyoto Protocol
>
> 8/CMP.8   Methodology for the collection of international transaction log fees in the biennium 2014–2015
>
> 9/CMP.8   Proposal from Kazakhstan to amend Annex B to the Kyoto Protocol
>
> 10/CMP.8 Capacity-building under the Kyoto Protocol for developing countries
>
> 11/CMP.8 Capacity-building under the Kyoto Protocol for countries with economies in transition
>
> 12/CMP.8 Compliance Committee
>
> 13/CMP.8 Administrative, financial and institutional matters.
>
> Sources: UNFCCC/CMP (2013a,b).

Clean Development Mechanism that lists entities designated and professionally accredited by the Executive Board of the CDM to the CMP and is in relation to the "Korea Energy Management Corporation" (UNFCCC/CMP, 2013b, p. 12).

Warsaw Outcomes   The 19th session of the COP was held in Warsaw, Poland, in November 2013. Before referencing the key outcomes of COP-19, it is useful to focus briefly on the lead-up to Warsaw which was fraught with political tensions and resulted in the June 2013 meeting of the SBI, which was charged with laying the groundwork for the upcoming Warsaw COP-19 being unable to formally convene. As reported in Bloomberg News, the UN talks were set back by 6 months as the SBI failed to complete any work at 2 weeks of meetings in Bonn because Russia, Ukraine, and Belarus objected to the group's agenda (Vitelli and Nicola, 2013). Meanwhile, in an open letter dated 11 June 2013 which was disseminated in the same UN meeting, the former chair of the SBI, the former Argentine Ambassador Raul Estrada Oyuela, a veteran of climate change negotiations, requested that the current chair of the SBI from Poland resign immediately, stating:

> There are no precedents of the frustration of an important international meeting comparable to the one originated in your failure to conduct the work.... Of course there are also administrative costs as a consequence of your lack of success, but you are not going to be charged for the expenses.... In addition, your performance is the worst announcement for COP-19 to be held in your country" (Estrada-Oyuela, 2013).

While the rebuke was stingingly public, the reality was that 2 weeks of SBI negotiations came to a standstill, in spite of the travel costs associated with delegates who had come from all over the world, as well as the costs of convening the meeting. At issue was a procedural and legal concern relating to the decision-making of presiding officials at the conclusion of COP and CMPs where Russia, supported by others, wanted to be included in the provisional agenda of the SBI – this concern emanated from the way in which the Durban COP-18 had been gaveled to a close. Essentially, without agreement on the adoption of the agenda of the 38th session, the SBI could not meet to do its work (Third World Network, 2013).

Despite this shaky start, the Warsaw COP met as scheduled, and according to the UNFCCC, the key outcomes and developments of Warsaw COP-19 included (see the UNFCCC website):

- Governments agreed to communicate their respective contributions towards the universal agreement well in advance of the COP in Paris in 2015 and that the required MRV arrangements for domestic action had been finalized for implementation, thereby providing a solid foundation for the 2015 agreement.
- Further progress was also made in helping countries, especially the poorest, adapt to the impacts of climate change and build their own sustainable, clean energy futures.
- The GCF, planned to be a major channel of financing for developing world action, would be ready for capitalization in the second half of 2014.

- The rulebook for reducing emissions from deforestation and forest degradation was agreed, together with measures to bolster forest preservation and a results-based payment system to promote forest protection.
- Governments agreed on a mechanism to address loss and damage caused by long-term climate change impacts.
- COP-19 in Warsaw also provided a showcase for climate action by business, cities, regions, and civil society.

Stavins (2014a) has argued that the Warsaw COP outcomes are a cause for cautious optimism, citing for instance, the agreement that identified six components for the new ADP (mitigation, adaptation, finance, technology development and transfer, capacity-building, and transparency of action and support) and the agreement on a Warsaw Mechanism for loss and damage which mentions neither liability nor compensation and moves the topic for future consideration under the broad rubric of adaptation, but also notes that no progress was made on the issue of climate financing. But the Warsaw COP-19 was also witness to some new developments in terms of a walkout by environment and development NGOs. Sethi (2013) noted that environmental NGOs had always been an integral part of the climate talks, and that in a rare sign of frustration and solidarity, 800 NGO representatives staged a walkout from the climate negotiations. COP-19 was not without drama as 800 people representing NGOs returned their registration badges to the UN and left Poland's national stadium. In another unprecedented development, Marcin Korlec, the Polish Environment Minister, who was presiding over the talks, was sacked on November 21 in an move apparently meant to signal a push towards accelerated shale gas operations in Poland (Vidal and Harvey, 2013). In its final summary analysis of COP-19, ENB pointed out that the upcoming 2014 UN Climate Summit could infuse some vigor into the process of UN-led climate change negotiations, but the more sobering reminder of the current state of play in global climate change negotiations was that: "Ultimately, the question is if climate change will wait for the UNFCCC. Thus far, the evidence shows the UNFCCC is being left behind" (IISD/ENB, 2013, p. 29).

Boxes 2.10 and 2.11 list the decisions adopted by COP-19 and those adopted by CMP at its ninth session (CMP-9) in Warsaw, which, taken together, represent all the agreed outcomes emanating from Warsaw.

An examination of the UNFCCC report containing all of the 23 decisions taken by COP-19, and all 10 decisions of CMP-9 reveals not one single reference to "poverty", and not one single reference to "energy" (UNFCCC, 2014a,b).

The Warsaw outcomes are different from the outcomes of the preceding COPs which contained at least a minimal set of references to "energy" and

Box 2.10 Decisions adopted by the Warsaw COP-19

1/CP.19   Further advancing the Durban Platform

2/CP.19   Warsaw international mechanism for loss and damage associated with climate change impacts

3/CP.19   Long-term climate finance

4/CP.19   Report of the Green Climate Fund to the Conference of the Parties and guidance to the Green Climate Fund

5/CP.19   Arrangements between the Conference of the Parties and the Green Climate Fund

6/CP.19   Report of the Global Environment Facility to the Conference of the Parties and guidance to the Global Environment Facility

7/CP.19   Report of the Standing Committee on Finance to the Conference of the Parties

8/CP.19   Fifth review of the financial mechanism

9/CP.19   Work programme on results-based finance to progress the full implementation of the activities referred to in decision 1/CP.16, paragraph 70

10/CP.19   Coordination of support for the implementation of activities in relation to mitigation actions in the forest sector by developing countries, including institutional arrangements

11/CP.19   Modalities for national forest monitoring system

12/CP.19   The timing and the frequency of presentations of the summary of information on how all the safeguards referred to in decision 1/CP.16, appendix I, are being addressed and respected

13/CP.19   Guidelines and procedures for the technical assessment of submissions from Parties on proposed forest reference emission levels and/or forest reference level

14/CP.19   Modalities for measuring, reporting and verifying

15/CP.19   Addressing the drivers of deforestation and forest degradation

16/CP.19   Work of the Adaptation Committee

17/CP.19   Nairobi work programme on impacts, vulnerability and adaptation to climate change

18/CP.19   National adaptation plans

19/CP.19   Work of the Consultative Group of Experts on National Communications from Parties not included in Annex I to the Convention

20/CP.19   Composition, modalities and procedures of the team of technical experts under international consultation and analysis

21/CP.19   General guidelines for domestic measurement, reporting and verification of domestically supported nationally appropriate mitigation actions by developing country Parties

22/CP.19   Sixth national communications from Parties included in Annex I to the Convention

23/CP.19   Work programme on the revision of the guidelines for the review of biennial reports and national communications, including national inventory reviews, for developed country Parties

Sources: UNFCCC (2014a,b).

Box 2.11   Decisions adopted by the Conference of the Parties serving as the meeting of the Parties to the Kyoto Protocol at its ninth session (CMP-9)

1/CMP.9	Report of the Adaptation Fund Board
2/CMP.9	Second review of the Adaptation Fund
3/CMP.9	Guidance relating to the clean development mechanism
4/CMP.9	Review of the modalities and procedures for the clean development mechanism
5/CMP.9	Guidance on the implementation of Article 6 of the Kyoto Protocol
6/CMP.9	Guidance for reporting information on activities under Article 3, paragraphs 3 and 4,of the Kyoto Protocol
7/CMP.9	Modalities for expediting the establishment of eligibility for Parties included in Annex I with commitments for the second commitment period whose eligibility has not yet been established
8/CMP.9	Compliance Committee
9/CMP.9	Supplementary information incorporated in sixth national communications submitted in accordance with Article 7, paragraph 2, of the Kyoto Protocol
10/CMP.9	Programme budget for the biennium 2014–2015

Source: UNFCCC/CMP (2014).

"poverty". The fact that the examination of all the decisions taken by COP-19 and CMP-9 reveals that there is not one single mention of "energy", let alone "energy access" or "access to energy", and not one single mention of "poverty", let alone "poverty eradication" or "poverty reduction", has to be highlighted. The complete absence of any mention of energy issues and poverty reduction is quite evident, and should be contrasted with the minimal number of references to energy and poverty eradication in earlier COP outcomes. Given the importance of energy sector impacts on anthropogenic climate change, and the development impacts and linkages between poverty and climate change, this complete absence of any references to energy and poverty within the COP-19 agreed outcomes is worth underscoring.

Lima COP-20's Call for Climate Action   The Lima Conference held in the first 2 weeks of December included both the 20th session of the COP to the UNFCCC, and the 10th session of the COP serving as the Meeting of the Parties to the Kyoto Protocol (CMP-10). In addition, as in the case of preceding

COPs, the three subsidiary bodies of the UNFCCC –the SBSTA and the SBI – met for their 41st sessions, and the seventh part of the second session of the ADP (ADP 2-7) also met. The significance of the Lima COP is that it provided the last global UNFCCC negotiating forum to prepare a much-needed draft negotiating text that is anticipated to inform the scope and content of the 2015 COP-21 in Paris.

As highlighted previously, the one trend that was common and escalating in nature with regard to previous COP negotiations was the time taken to gavel the meeting to its conclusion, with the Lima intergovernmental process going 30 hours past the official deadline to arrive at a final agreement. According to a summary of COP-20 by the Center for Climate and Energy Solutions (2014), COP-20 managed to "hammer out a modest set of procedural steps, and made no real progress on the larger issues looming as they work toward a new global climate agreement next year in Paris". The report went on to note that the COP began with a "sense of momentum" resulting from the announcements of the close to $10 billion in pledges to the newly created GCF and the joint US–China proposal on their respective post-2020 emission targets, but then got "quickly bogged down" as parties moved away from their discussion on the draft elements of an agreement to "haggle over the more immediate issues of how their intended contributions to the Paris agreement are to be submitted to and weighed" (p. 1). However, according to Stavins (2014b), the Lima COP represented both a "classic compromise" between developed and developing countries, and "something of a breakthrough after twenty years of difficult climate negotiations", because the agreement on the submission of INDCs from "all parties" should be viewed as "a significant departure from the past two decades of international climate policy" (Stavins, 2014b). Meanwhile, the ENB summary report of the Lima COP noted that the issues such as "differentiation" of parties' responsibilities and the role of the UNFCC were keenly debated, especially the distinctions between following established principles of "common but differentiated responsibilities and equity " or between the "principle of common but differentiated responsibility and "respective capabilities in line with varying national circumstances" in regards to the INDCs and draft elements reflected in LCCA. The ENB summary report discussed the push back from developing countries against the "mitigation centric" approach, and towards a parity approach between mitigation and adaptation in the proposed INDCs; and stated that final agreement "arguably, shifts the wall of differentiation", because the ADP's commitment to secure a 2015 agreement reflects "the principle of common but differentiated responsibility and respective capabilities in light of different national circumstances", and this "formulation" tended to favor "a subjective interpretation of differentiation" (ENB, 2015, p. 44).

There may be disagreement as to whether the LCCA represents incremental progress that will result in an ambitious climate agreement in Paris, or whether it represents a worrisome slowing down that will result in either a stalled or watered-down agreement in Paris. The bottom line is that there is broad global consensus on the need for urgent climate action but no one can predict the exact level of ambition and scope of climate action that will finally emerge in 2015. As discussed in Chapter 1, the LCCA's emphasis on the submission of INDCs by all Parties is not only a clear indication of the role of the INDCs in any future climate change agreement, but also a signal for a potentially more inclusive approach to climate change mitigation for the first time in the history of the UNFCCC process. Unlike the 1992 UNFCCC, which clearly distinguished between the roles and responsibilities of Annex I and non-Annex I Parties, all Parties are encouraged to submit INDCs, but whether and how individual countries and groups of countries, especially those constituting the major emitters, will commit to time-bound climate mitigation remains to be seen. However, as discussed in Chapter 1, the absence of a concrete agreement as to the basic and common template for the information to be communicated, let alone an agreement on the baseline years and accounting processes to be used for INDCs, and any specific guidance as to the exact process by which INDCs will be assessed and by whom, leaves it unclear as to the diversity of information that future INDC submission may or may not include, and whether the submission of proposed INDCs will even lend themselves to any sort of future assessment or review.

As discussed in Chapter 1, the significant contribution of the Lima conference is the Annex to Decision 1, entitled "Elements for a draft negotiating text", because it is this Annex that provides the basic draft that is expected to be used by negotiators to come with a final draft negotiations text in May 2015, which in turn will be the basis of negotiations at the Paris COP-21. The stakes are extremely high for this Paris COP because the 2015 deadline is one that has been much anticipated and any agreement arrived at in Paris in December 2015 will have to be brought into the broader UN-led quest for a shared post-2015 development agenda and the 2015 agreement on sustainable development goals. The astonishing proposal to establish a tax on the oil exports from developing to developed countries, which was highlighted in Chapter 1, is particularly worth reiterating at this juncture, precisely because such a proposal has huge socio-economic development implications and costs that appear to be selectively levied against developing countries only, and also because such a proposal contradicts the long-stated, previously agreed UNFCCC preambular recognition that "especially developing countries, need access to resources" and that "their energy consumption will need to grow", as well as, Articles 4.8 and 4.10 (commitments), which focuses expressly on the needs of developing country Parties, including "countries whose economies are highly dependent on income generated from the production, processing and export/ and/or on consumption of fossil fuels" (IUCC, 1992, pp. 4 and 11). It will be absolutely critical to see whether this idea of a tax of oil export from developing to developed countries is carried over to any future climate agreement emanating from COP-21, because it has the potential to send a shock wave in terms of energy markets and prospects for oil exports in developing countries. In stark contrast with this radical proposal, which has the potential to seriously impact on the energy and socio-economic development nexus in developing countries, and which has no analog anywhere in terms of any of the global climate change agreed outcomes for the past 20 years, an examination of the Lima outcomes confirms the 20-year trend towards a paucity of concrete linkages between energy access for the poor and climate change objectives.

An examination of the Lima outcomes, including those resulting from COP-10 contained in Box 2.12 and those resulting from CMP-10 contained in Box 2.13 reveal that references to the nexus between energy for sustainable development objectives and climate change, in particular references that could potentially link goals and targets on energy access for the poor with goals and targets on climate change objectives, are entirely missing.

A detailed examination of Decisions 1–24 of the Lima COP reveals that none of the decisions contain any mention of the issue of "access to energy" or "energy access for the poor" (UNFCCC, 2015a–c). Unlike the Warsaw outcomes however, there are four references in total to "poverty" in all of the Lima COP decisions, and all four references are contained in the Annex to Decision 1 entitled, "Elements for a draft negotiating text", which will be used as the basis for negotiators to come, with the May 2015 version of the text forwarded for final negotiations at COP-21 (UNFCCC, 2015a).

The first reference to "poverty" is contained in Option D of the preambular draft section and is a reaffirmation to "take full account of the legitimate priority needs of developing countries and their right to equitable access to sustainable development and for achieving economic growth and the eradication of poverty" (UNFCCC, 2015a, p. 7). The next two references to "poverty" can be found in paragraph 2 of Section C (general/objective), which calls on all Parties:

> …to strive to achieve low greenhouse gas climate-resilient economies and societies, on the basis of equity and in accordance with their historical responsibilities… in order to achieve sustainable development, poverty eradication… taking fully into account the historical responsibility of developed country Parties and their leadership in combating climate change and the adverse effects thereof, and bearing in mind that economic and social development and poverty eradication are the first and overriding priorities of developing country Parties (UNFCCC, 2015a, p. 8).

The final mention of poverty can be found in draft Option C of paragraph 76 (transparency/consultative process), which discusses the need to "facilitate understanding of the diversity, barriers and needs, the enhanced actions

Box 2.12 Decisions adopted by the 20th Conference of the Parties (COP-20) in Lima

1/CP.20   Lima Call for Climate Action

2/CP.20   Warsaw International Mechanism for Loss and Damage associated with Climate Change Impacts

3/CP.20   National adaptation plans

4/CP.20   Report of the Adaptation Committee

5/CP.20   Long-term climate finance

6/CP.20   Report of the Standing Committee on Finance

7/CP.20   Report of the Green Climate Fund to the Conference of the Parties and guidance to the Green Climate Fund

8/CP.20   Report of the Global Environment Facility to the Conference of the Parties and guidance to the Global Environment Facility

9/CP.20   Fifth review of the Financial Mechanism

10/CP.20   Further guidance to the Least Developed Countries Fund

11/CP.20   Methodologies for the reporting of financial information by Parties included in Annex I to the Convention

12/CP.20   Fifth Assessment Report of the Intergovernmental Panel on Climate Change

13/CP.20   Guidelines for the technical review of information reported under the Convention related to greenhouse gas inventories, biennial reports and national communications by Parties included in Annex I to the Convention

14/CP.20   Training programme for review experts for the technical review of greenhouse gas inventories of Parties included in Annex I to the Convention

15/CP.20   Training programme for review experts for the technical review of biennial reports and national communications of Parties included in Annex I to the Convention

16/CP.20   Joint annual report of the Technology Executive Committee and the Climate Technology Centre and Network for 2013

17/CP.20   Joint annual report of the Technology Executive Committee and the Climate Technology Centre and Network for 2014

18/CP.20   Forum and work programme on the impact of the implementation of response measures

19/CP.20   Parties included in Annex I to the Convention whose special circumstances are recognized by the Conference of the Parties

20/CP.20   Lima work programme on gender

21/CP.20   The Lima Ministerial Declaration on Education and Awareness-raising

22/CP.20   Administrative, financial and institutional matters

23/CP.20   Revisions to the financial procedures for the Conference of the Parties, its subsidiary bodies and the secretariat

24/CP.20   Dates and venues of future sessions

Sources: UNFCCC (2015a–c).

> ### Box 2.13   Decisions adopted by the Conference of the Parties serving as the meeting of the Parties to the Kyoto Protocol at its 10th session (CMP-10)
>
> 1/CMP.10   Report of the Adaptation Fund Board
> 2/CMP.10   Second review of the Adaptation Fund
> 3/CMP.10   Date of the completion of the expert review process under Article 8 of the Kyoto Protocol for the first commitment period
> 4/CMP.10   Guidance relating to the clean development mechanism
> 5/CMP.10   Guidance on the implementation of Article 6 of the Kyoto Protocol
> 6/CMP.10   Synergy relating to accreditation under the mechanisms of the Kyoto Protocol
> 7/CMP.10   Outcome of the work programme on modalities and procedures for possible additional land use, land-use change and forestry activities under the clean development mechanism
> 8 /CMP.10   Administrative, financial and institutional matters
>
> Source: UNFCCC/CMP (2015).

undertaken by developing country Parties, bearing in mind their first and overriding priority of economic and social development and poverty eradication" (UNFCCC, 2015a, p. 35). What is clear from this examination is that poverty eradication is quite specifically referenced as a key priority for developing countries, which makes it even harder to explain the complete absence of concrete linkages and guidance on energy access for the poor and climate change objectives.

A search of Decision 1 and its Annex – the LCCA – indicates that "energy" is referenced a total of seven times in this very important and critical outcome decision of the Lima COP. The first two references are contained in a footnote to the agreed portion of the LCCA, which discusses the fact that in 2014, technical expert meetings convened under the aegis of the ADP were undertaken in areas of "renewable energy, energy efficiency, land-use change and forestry (including REDD-plus), urban environments, carbon dioxide capture use and storage and non-CO_2 greenhouse gases" (UNFCCC, 2015a, p. 4). The next two references are contained in the preambular section of the Annex entitled "Elements for a draft negotiating text", and can be found under Option (d) of the preambular section, which borrowed exact language from the 1992 UNFCCC preambular section cited previously, and reaffirmed that:

> ...all developing countries need access to the resources required to achieve sustainable social and economic development and that, in order for developing countries to progress towards that goal, their energy consumption will need to grow, taking into account the opportunities for achieving greater energy efficiency and for reducing greenhouse gas emissions, including through the application of new technologies on terms which make such an application economically and socially beneficial (UNFCCC, 2015a, p. 7).

It should be noted that the proposal for the tax on oil exports from developing countries can be seen as contravening the need for developing countries to access their energy resources.

The final three references to "energy" are all found in section G of the Annex dealing with finance. The first reference calls for the mobilization and support of finance that can "support the integration of climate objectives into other policy-relevant areas and activities such as energy, agriculture, planning and transport" (UNFCCC, 2015a, p. 19), while the remaining two references are contained in paragraph 53.1, which calls for "an international renewable energy and energy efficiency bond facility to be established" and which is listed just below the call for a tax on oil exports from developing countries to developed countries (UNFCCC, 2015a, p. 24).

Of the remaining Decisions 2–24, only Decision 14, entitled "Training programme for review experts for the technical review of greenhouse gas inventories of Parties included in Annex I to the Convention", contains one reference to "energy". "Energy" is mentioned in connection to the description of a basic course for the review of GHG inventories of Annex I Parties (developed) to the UNFCCC and focuses on the role of inventory guidance provided by the IPCCC in relation to a number of sectors, including energy (UNFCCC, 2015c, p. 27).

In addition to the COP-20 decisions listed earlier, the Lima Conference resulted in a series of decisions emanating from the 10th session of the CMP (CMP-10) which are listed in Box 2.13.

There are no references whatsoever to "poverty" in any of the decisions emanating from CMP-10 (UNFCCC/CMP, 2015). There are only two references to "energy" in all eight decisions emanating from the Lima CMP-10. As in the case of the review of the Doha CMP-8, these two references to energy are both contained in an Annex related to Decision 4 (Guidance to the Clean Development Mechanism) of CMP-10 and reference the "Korea Energy Management Corporation", which is listed twice concurrently within the list of "Entities accredited and provisionally designated by the Executive Board of the clean development mechanism" (UNFCC/CMP, 2015, p. 19).

2.4   The absence of concrete references to "energy access for the poor" in key agreed global climate outputs: A puzzling disconnect in the lead-up to 2015

In spite of the central role of fossil fuel energy-related GHGs in anthropogenic climate change, concrete references and specific guidance on energy are largely absent in key global climate change outcomes. As discussed in Chapter 1, the lack of access to sustainable energy services has been well recognized in impacting negatively on the lives of the poor. Programmatic references and policy guidance on energy issues in general, and on energy access for the poor in particular, are, however, missing from key agreed global climate change decisions. The objective of this chapter was to review key global outcomes agreed to within the context of over 20 years of intergovernmental negotiations in terms of the nexus between energy and climate change – more specifically, the nexus between energy access for the poor and climate change. The detailed examination of key climate change outcomes, starting from the 1992 UNFCCC and going on to the 2014 COP-19 Warsaw outcomes, provides clear evidence of the absence of references to energy access for the poor or increasing access to sustainable energy services for developing countries, as well as a surprisingly minimal amount of linkages between poverty reduction and climate change objectives.

Concrete policy guidance and programmatic references to energy access and poverty reduction in key global climate change outcomes negotiated under the aegis of the UN are crucial to the broader global quest for a low-carbon, sustainable future. The absence of such references and linkages between climate change and energy access for the poor is most detrimental to poor communities and countries, and is harder to reconcile in terms of the broad consensus and global priority accorded to poverty reduction within the context of the UN. The double burden of being "energy poor" and "climate vulnerable" has been recognized by globally relevant reports and organizations, yet somehow is not borne out in the actual agreed record on key global climate change outcomes.

This absence of explicit linkages and policy guidance on energy access in key global climate change outcomes over the course of two decades does not appear to encourage effective synergies between poverty reduction and climate change goals. And so for instance, the lack of a shared framework for action in terms of increasing energy access for the poor and addressing SLCPs inhibits integrated responses and actions. As such, this may be construed as wasteful and untimely in terms of delivering complementary and cost-effective efforts related to intrinsically linked development objectives of increasing access to sustainable energy services for the poor and addressing climate change. Additionally, the policy linkages between climate change and renewable energy have not been made explicit in recent agreed outcomes of the global climate change negotiations, even though a recent IPCC Special Report on Renewable Energy Sources and Climate Change Mitigation (SRREN), produced by a global team of 120 researchers, which reviewed six of the most important renewable energy technologies (bioenergy, direct solar energy, geothermal energy, hydropower, ocean energy, and wind energy), recognized that close to 80% of the world's energy supply could be met by renewables by mid-century if backed by the right enabling public policies.

To date, global climate change negotiations have generated a vast array of specialized committees and working groups making recommendations and putting forward draft text on a dizzying number of climate change-related issues. But, as the long-drawn-out concluding plenaries of recent COPs, from 2009 onwards, have demonstrated, securing a global agreement by consensus is tedious, because it only takes a small handful of countries to block agreement. The climate change negotiations have also revealed new rifts and negotiating blocs amongst groups of countries, which make a consensus-based agreement harder to achieve. The big question is whether the lack of concrete time-bound outcomes in Copenhagen and Cancun signifies a fundamental shift in framework of climate change negotiations from consensus- and compliance-based mandatory emissions caps towards less multilateral, non-mandatory, and voluntary national measures. What is equally important to note is that the process of negotiating to draft the Copenhagen Accord language occurred amongst a select group of countries, outside of the multilateral climate change negotiations and behind closed doors.

The crafting of the Accord brought a new constellation of countries, including the BASIC countries. But, the emergence of a plethora of groups and negotiating blocs during the course of more than 20 years of global climate change negotiations held in a diverse array of countries across the world has not resulted in any immediate clarity or urgency regarding the securing of a comprehensive, global consensus-based climate change mitigation agreement.

The current framework of the UN-led negotiations has long been seen as allowing for the broadest possible representation of countries, with each country having an equal voice regardless of size. It provides a forum that is aimed at being democratic and open in terms of participation and inclusion to state and non-state actors, but there is also a growing recognition of the challenges associated with securing time-bound, consensus-based agreements, precisely because of the sheer number of actors involved and the increasingly complex and lengthy nature of the negotiations. However, the broad push towards a consensus-based approach in climate change is expected to culminate in the much anticipated climate deal at the COP-21 in Paris in 2015 (Aldy and Stavins, 2009).

Given the central role of energy in driving anthropogenic climate change, and the overall importance of increasing access to sustainable energy as a means of reducing poverty and addressing sustainable development, it is puzzling that there are no references whatsoever to sustainable energy access for the poor within agreed global climate change outputs examined earlier in the chapter. The absence of detailed policy and programmatic references to energy and poverty reduction in a long series of climate change negotiated outcomes, combined with the fact that the UN has long embarked on a separate set of negotiations on energy for sustainable development objectives, is worth underscoring. Reluctance to adopt concrete emissions reductions and acceptance of more voluntary, non-mandatory goals appear increasingly to be the norm expressed by key aggregate emitters. With the 2015 deadline fast approaching to secure an international framework agreement aimed at achieving specific global GHG emissions targets, it is still not clear what the exact terms of the mitigation agreement will be, and which countries will clearly commit to a legally binding agreement. The elements for a draft text contained in the LCCA are supposed to provide the rubric for the anticipated climate deal expected to be agreed to by COP-21 in 2015, but it remains to be seen whether intrinsically linked development challenges like energy and poverty reduction will once again be absent from any future climate agreement, and whether the nexus between energy access for the poor and climate change will once again be ignored.

In conclusion, a series of questions related to the energy access–climate change relationship can be raised with the attendant hope that answers to these questions will be forthcoming in terms of agreed intergovernmental outputs in the coming year:

- Will there be a clear set of goals and targets related to climate change that will be agreed to at the upcoming Paris COP-21 in 2015, and what will the anticipated 2015 climate change-related sustainable development goal finally entail?
- Why is there a lack of consistent concrete measures and targets focused on energy access for the poor, including the role of SLCPs, and financing for sustainable energy services for the poor within the context of agreed global climate change outcomes?
- Will the linkages between increasing access to sustainable energy services for the poor and climate change, in particular the role of SLCPs, be concretely addressed in any comprehensive climate change agreement anticipated at COP-21 or in the 2015 global sustainable development goals?
- Will there be any opportunities for a nexus between agreed climate change goals and targets and agreed energy access goals and targets that are the subject of two different intergovernmental negotiating silos, particularly in light of the UN quest for a shared post-2015 development agenda?

References

Adler, E. and Haas, P. (1992) Conclusion: epistemic communities, world order and the creation of a reflective research program. *International Organisation*, 46(1), 367–390.

Aldy, J. and Stavins, R. (eds) (2009) *Post-Kyoto International Climate Policy: Summary for Policymakers*. Cambridge: Cambridge University Press.

Babiker, M., Reilly, J. and Jacoby, H.D. (2000) The Kyoto Protocol and developing countries. *Energy Policy*, 28(8), 525–536.

Bernauer, T. (1995) The effect of international environmental institutions: how we might learn more. *International Organization*, 49, 351–377.

Bodansky, D. (2011) *Governing Climate Engineering: Scenarios for Analysis*. Cambridge: Harvard Project on Climate Agreements, Discussion Paper, 11–47.

Bodansky, D. (2012) *The Durban Platform: Goals and Options*. Cambridge: Harvard Project on Climate Agreements, Viewpoints. http://belfercenter.ksg.harvard.edu/files/bodansky_durban2_vp.pdf [accessed May 11, 2015].

Bodansky, D. and Diringer, E. (2010) *The Evolution of Multilateral Regimes: Implications for Climate Change*. Arlington: Pew Center on Global Climate Change.

Carpenter, C. (2012) *Taking Stock of Durban: Review of Key Outcomes and the Road Ahead*. New York: UNDP/EEG.

Chasek, P. and Wagner, L. (2012) *The Roads from Rio: Lessons Learned from Twenty Years of Multilateral Environmental Negotiations*. New York: Routledge.

Center for Climate and Energy Solutions (2014) Outcomes of the U.N. Climate Change Conference in Lima. http://www.c2es.org/docUploads/cop-20-summary.pdf [accessed January 12, 2014].

Cherian, A. (1997) Energy policies, liberalization and the framing of climate change policies in India. *Doctoral Dissertation*, University of Massachusetts, Amherst.

Cherian, A. (2007) Linkages between biodiversity conservation and global climate change in small island developing states. *Natural Resources Forum*, 31(2), 128–131.

Cherian, A. (2009) *Bridging the Divide Between Poverty Reduction and Climate Change through Sustainable and Innovative Energy Technologies*. New York: UNDP Expert Paper.

Ciscar, J.C. and Soria, A. (2002) Prospective analysis of beyond Kyoto climate policy: a sequential game framework. *Energy Policy*, 30(15), 1327–1335.

Climate Change Secretariat (1998) *The Kyoto Protocol to the Convention on Climate Change*. Geneva: UNEP/IUC.

Depledge, J. (2005) *The Organization of Global Negotiations: Constructing the Climate Change Regime*. London: Earthscan.

Dodds, F., Laguna-Celis, J. and Thompson, L. (2014) *From Rio+20 to a New Development Agenda*. New York: Routledge.

Downs, Anthony (1972) Up and down with ecology-the issue-attention cycle. *Public Interest*, 2(28), 38–50.

ENB (2009) Analysis of COP 15/CMP 5. *Earth Negotiations Bulletin*, 12(459). IISD/ENB.

ENB (2010) Summary of the Cancun Climate Change Conference. *Earth Negotiations Bulletin*, 12(498). www.iisd.ca/vol12/enb12498e.html [accessed May 11, 2015].

ENB (2012) Summary of the Doha Climate Change Conference. *Earth Negotiations Bulletin*, 12(567). IISD/ENB. http://www.iisd.ca/vol12/enb12567e.html [accessed on Feb 26, 2014].

ENB (2013) Summary of the Warsaw Climate Change Conference. *Earth Negotiations Bulletin*, 12(594). http://www.iisd.ca/vol12/enb12594e.html [accessed March 11, 2014].

ENB (2014) Summary of the Lima Climate Change Conference. *Earth Negotiations Bulletin*, 12(619). http://www.iisd.ca/vol12/enb12619e.html [accessed on January 14, 2015].

Estrada-Oyuela, R. (2013) Open Letter to Mr Tomasz Chruszczow Chairman of the SBI 38, June 11, 2013. Buenos Aires.

Flavin, C. (1997) Financing a new energy economy: a key role for the World Bank in playing God with climate. *World Watch*, 10(6), 25–35.

Gillis, J. (2014) "U.N. Says Lag in Confronting Climate Woes Will be Costly". *New York Times*, January 16, 2014.

Gore, Al. (2006) *An Inconvenient Truth: The Planetary Emergency of Global Warming and What We Can Do About It*. New York: Rodale.

Grubb, M. and Anderson, D. (eds) (1995) *The Emerging International Regime for Climate Change: Structures & Options After Berlin*. Chatham: Royal Institute of International Affairs.

Grubb, M. and Yamin, F. (2001) Climatic collapse at the Hague: what happened, why, and where do we go from here? *International Affairs*, 77(2), 261–276.

Gupta., J (2000) North-south aspects of the climate change issue: towards a negotiating theory and strategy for developing countries. *International Journal of Sustainable Development*, 3(2), 115–135.

Haas, P. (1992) Introduction: epistemic communities and international policy coordination. *International Organization*, 46(1).

Haas, P. (2012) The political economy of ecology: prospects for transforming the world economy at Rio Plus 20. *Global Policy*, 3(1).

Harvey, F. (2012) "Doha climate change deal clears way for 'damage aid' to poor nations". *The Observer,* December 8, 2012.

Hawkins, A. (1993) Contested ground: international environmentalism and global climate change. In: Lipschutz, R. and Conca, K., eds. *The State and Social Power in Global Environmental Politics*. New York: Columbia University Press.

Hoffmann, M. (2011) *Climate Governance at the Crossroads: Experimenting with a Global Response after Kyoto*. New York: Oxford University Press.

Houghton, J.T. (2009) *Global Warming: The Complete Briefing*. Cambridge: Cambridge University Press.

Hultman, N. (2012) *The Durban Platform*. Washington, DC: Brookings Institute. http://www.brookings.edu/research/opinions/2011/12/12-durban-platform-hultman [accessed May 11, 2015].

Hurrell, A. and Kingsbury, B. (eds) (1992) *International Politics of the Environment*. Oxford: Clarendon Press.

Hyder, T.O. (1994) Looking back to see forwar. In: Minzter, I. and Leonard, J.A., eds. *Negotiating Climate Change: The Inside Story of the Rio Convention*. Cambridge: Cambridge University Press.

IISD/ENB (2009) Brief Analysis of the Copenhagen Climate Change Conference. IISD, pp. 1–7, Available at http://www.iisd.org/pdf/2009/enb_copenhagen_commentary.pdf [accessed February 26, 2014].

IISD Reporting Services (2009) UN Climate Conference Concludes by Taking Note of the Copenhagen Accord. ISSD, December 19, 2009. http://climate-l.iisd.org/news/un-climate-change-conference-concludes-by-taking-note-of-the-%E2%80%9Ccopenhagen-accord%E2%80%9D/ [accessed on March 12, 2014].

IUCC (1992) *United Nations Framework Convention on Climate Change.* Geneva: UNEP/WMO IUCC.

IRENA (2012) *IRENA Handbook on Renewable Energy Nationally Appropriate Mitigation Actions (NAMAs) for Policy Makers and project Developers.* Abu Dhabi: IRENA.

Keohane, R. and Levy, M. (1996) *Institutions for Environmental Aid.* Cambridge: MIT Press.

Khan, M. (2014) *Towards a Binding Climate Change Adaptation Regime: A Proposed Framework.* New York: Routledge.

Levy, M., Keohane, R. and Haas, P. (eds) (1993) *Institutions for the Earth: Sources of Effective International Environmental Protection.* Cambridge: MIT Press.

Levy, M., Young, O. and Zurn, M. (1994) *The Study of International Regimes.* Laxenburg: IIASA.

Litfin, K. (1993) Ecoregimes: playing tug of war with the nation-state. In: Lipschutz, R. and Conca, K., eds. *The State and Social Power in Global Environmental Politics.* New York: Columbia University Press.

Lütken, S., Fenhann, J., Hinostroza, M. *et al.* (2011) *Low Carbon Development Strategies: A Primer on Framing Nationally Appropriate Mitigation Actions (NAMAs) in Developing Countries.* Denmark: UNEP Risoe.

Mattoo, A. and Subramanian, A. (2013) *Greenprint: A New Approach to Cooperation on Climate Change.* Washington D.C., Brookings Institute Press.

Ott, H. (2001) Climate Change an important foreign policy issue. *International Affairs*, 77(2), 277–296.

Murray, J. (2010) BASIC Countries to meet ahead of crucial Copenhagen accord deadline. *The Guardian*, January 12, 2010. http://www.theguardian.com/environment/2010/jan/12/copenhagen-climate-change [accessed August 20, 2014].

Paterson, M. and Grubb, M. (1992) The international politics of climate change. *International Affairs*, 68(2), 293–310.

Paterson, M. and Grubb, M. (eds) (1996) *Sharing the Effort: Options for Differentiating Commitments on Climate Change.* Washington, DC: Brooking Institute Press.

Quarless, D. (2007) Addressing the vulnerability of SIDS. *Natural Resources Forum*, 31(2), 99–101.

Ravindranath, N. and Sayathe, J. (2002) *Climate Change and Developing Countries.* Dordrecht: Kluwer Academic Publishers.

Ritter, K. (2012) UN Climate Talks in Doha, Qatar face multiple challenges. *AP/Huffington Post*. November 25, 2012. http://www.huffingtonpost.com/2012/11/25/2012-un-climate-talks-qatar_n_2188048.html [accessed May 11, 2015].

Roberts, T. and Parks, B. (2007) *Climate of Injustice: Global Inequality, North-South Politics, and Climate Policy.* Cambridge: MIT Press.

Roberts, T. and Edwards, G. (2012) *A New Latin American Climate Negotiating Group: The Greenest shoots in the Doha Desert*. Washington, DC: Brookings Institute. Blog: http://www.brookings.edu/blogs/up-front/posts/2012/12/12-latin-america-climate-roberts [accessed May 11, 2015].

Rosa, L.P. and Munasinghe, M. (eds) (2002) *Ethics, Equity and International Negotiations on Climate Change*. Cheltenham: Edward Elgar Publishing.

Sethi, N. (2013) NGOs walk out of Warsaw Talks. *The Hindu*, November 22, 2013. http://www.thehindu.com/sci-tech/energy-and-environment/ngos-walk-out-of-warsaw-talks/article5376547.ece [accessed May 11, 2015].

Sharma, S. and Desgain, D. (2013) *Understanding the Concept of Nationally Appropriate Mitigation Actions*. Denmark: UNEP Risoe.

Sjöstedt, G. and Penetrante, A. eds. (2013) *Climate Change Negotiations: A Guide to Resolving Disputes and Facilitating Multilateral Cooperation*. New York: Routledge.

Soltau, F. (2009) *Fairness in International Climate Change Law and Policy*. Cambridge: Cambridge University Press.

Stavins, R. (2010) *What Happened (and Why): An Assessment of the Cancun Agreements*. Cambridge: Harvard Belfer Center. http://www.robertstavinsblog.org/2010/12/13/successful-outcome-of-climate-negotiations-in-cancun/ [accessed January 30, 2014].

Stavins, R. (2014a) The Warsaw climate negotiations and reasons for cautious optimism. *Huffington Post*, January 10, 2014. http://www.huffingtonpost.com/robert-stavins/the-warsaw-climate-negoti_b_4577321.html [accessed March 12, 2014].

Stavins, R. (2014b) *Assessing the Outcome of the Lima Climate Talks*. Cambridge: Harvard Belfer Center. http://www.robertstavinsblog.org/2014/12/14/assessing-the-outcome-of-the-lima-climate-talks/ [accessed January 15, 2015].

TERI (1992) *TERI Information Service on Global Warming (TISGLOW)*, 3(2), 60–69. New Delhi: TERI.

Toyne, P. (2011) What the Durban outcome means for business. *The Guardian*, December 12, 2011. http://www.theguardian.com/sustainable-business/blog/durban-cop17-outcome. [accessed May 11, 2015].

Third World Network (2013) UNFCCC Body in Crisis: unable to do work. *TWN Bonn News Update*, 17, June 12, 2013.

UN/CDP (2007) *International Development Agenda and the Climate Change Challenge*. New York: UN.

UNFCCC (2008) Report of the Conference of the Parties on its thirteenth session, held in Bali from 3 to 15 December 2007. Action taken by the Conference of the Parties at its thirteenth session: Addendum 1. FCCC/CP/2007/6/Add.1. http://unfccc.int/documentation/documents/advanced_search/items/6911.php?priref=600004671#beg [accessed January 28, 2014].

UNFCCC (2009) Draft decision -/CP.15: Proposal by the President-Copenhagen Accord FCCC/CP/2009/L.7. http://maindb.unfccc.int/library/view_pdf.pl?url=http://unfccc.int/resource/docs/2009/cop15/eng/l07.pdf [accessed January 12, 2015].

UNFCCC (2010) Report of the Conference of the Parties on its fifteenth session, held in Copenhagen from 7-19 December, 2009. Decisions adopted by the Conference of the Parties: Addendum.1. FCCC/CP/2009/11/Add.1. http://unfccc.int/resource/docs/2009/cop15/eng/11a01.pdf [accessed January 12, 2015].

UNFCCC (2011a) Report of the Conference of the Parties on its sixteenth Session held in Cancun from 29 November to 10 December, 2010. Decisions adopted by the Conference of the Parties: Addendum 1. FCCC/CP/2010/7/Add.1. Available at http://unfccc.int/resource/docs/2010/cop16/eng/07a01.pdf#page=2 [accessed January 12, 2015].

UNFCCC (2011b) Report of the Conference of the Parties on its sixteenth session, held in Cancun from 29 November to 10 December, 2010. Decisions adopted by the Conference of the Parties: Addendum 2 FCCC/CP/2010/7/Add.2. t http://unfccc.int/resource/docs/2010/cop16/eng/07a02.pdf [accessed January 12, 2015].

UNFCCC (2012a) Report of the Conference of the Parties on its seventeenth session, held in Durban from 28 November to 11 December, 2011. Decisions adopted by the COP: Addendum1FCCC/CP/2011/9/Add.1. http://unfccc.int/resource/docs/2011/cop17/eng/09a01.pdf [accessed January 13, 2015].

UNFCCC (2012b) Report of the Conference of the Parties on its seventeenth session, held in Durban from 28 November to 11 December 2011. Decisions adopted by the Conference of the Parties: Addendum 2. FCCC/CP/2011/9/Add.2 http://unfccc.int/resource/docs/2011/cop17/eng/09a02.pdf [accessed January 13, 2015].

UNFCCC (2013a) Report of the Conference of the Parties on its eighteenth session, held in Doha from 26 November to 8 December 2012. Decisions adopted by the Conference of the Parties: Addendum 1. FCCC/CP/2012/8/Add.1 http://unfccc.int/resource/docs/2012/cop18/eng/08a01.pdf [accessed January 13, 2015].

UNFCCC (2013b) Report of the Conference of the Parties on its eighteenth session, held in Doha from 26 November to 8 December 2012. Decisions adopted by the Conference of the Parties: Addendum 2. FCCC/CP/2012/8/Add.2. http://unfccc.int/resource/docs/2012/cop18/eng/08a02.pdf [accessed January 13, 2015].

UNFCCC (2013c) Report of the Conference of the Parties on its eighteenth session, held in Doha from 26 November to 8 December 2012. Decisions adopted by the Conference of the Parties: Addendum 3. FCCC/CP/2012/8/Add.3 http://unfccc.int/resource/docs/2012/cop18/eng/08a03.pdf [accessed January 13, 2015].

UNFCCC (2014a) Report of the Conference of the Parties on its nineteenth session held in Warsaw from 11 to 23 November 2013. Decisions adopted by the Conference of the Parties: Addendum 1.FCCC/CP/2013/10 http://unfccc.int/resource/docs/2013/cop19/eng/10.pdf [accessed January 14, 2015].

UNFCCC (2014b) Report of the Conference of the Parties on its nineteenth session held in Warsaw from 11 to 23 November 2013. Decisions adopted by the Conference of the Parties: Addendum 2. FCCC/CP/2013/10/Add.2/Rev.1. http://unfccc.int/resource/docs/2013/cop19/eng/10a02r01.pdf [accessed January 14, 2015].

UNFCCC (2015a) Report of the Conference of the Parties on its twentieth session held in Lima from 1 to 14 December 2014. Decisions adopted by the Conference of the Parties: Addendum 1 FCCC/CP/2014/10/Add.1. http://unfccc.int/resource/docs/2014/cop20/eng/10a01.pdf [accessed February 20, 2015].

UNFCCC (2015b) Report of the Conference of the Parties on its twentieth session held in Lima from 1 to 14 December 2014. Decisions adopted by the Conference of the Parties: Addendum 2. FCCC/CP/2014/10/ Add.2. http://unfccc.int/resource/docs/2014/cop20/eng/10a02.pdf [accessed February 20, 2015].

UNFCCC (2015c) Report of Conference of the Parties on its twentieth session held in Lima. Decisions Adopted by the Conferences of the Parties: Addendum 3.

FCCC/CP/2014/10/Add.3.   http://unfccc.int/resource/docs/2014/cop20/eng/10a03.pdf [accessed February 20, 2015].

UNFCCC/CMP (2011a) Report of the Conference of the Parties serving as the meeting of the Parties to the Kyoto Protocol on its sixth session. Action taken by the Conference of the Parties serving as the meeting of the Parties to the Kyoto Protocol: Addendum 1 FCCC/KP/CMP/2010/12/Add.1 http://unfccc.int/resource/docs/2010/cmp6/eng/12a01.pdf [accessed January 12, 2015].

UNFCCC/CMP (2011b) Report of the Conference of the Parties serving as the meeting of the Parties to the Kyoto Protocol on its sixth Session. Action taken by the Conference of the Parties serving as the meeting of the Parties to the Kyoto Protocol: Addendum 2. FCCC/KP/CMP/2010/12/Add.2. http://unfccc.int/resource/docs/2010/cmp6/eng/12a02.pdf [accessed January 12, 2015].

UNFCCC/CMP (2012a) Report of the Conference of the Parties serving as the meeting of the Parties to the Kyoto Protocol on its seventh session. Action taken by the Conference of the Parties serving as the meeting of the Parties to the Kyoto Protocol: Addendum1. FCCC/KP/CMP/2011/10/Add.1. http://unfccc.int/resource/docs/2011/cmp7/eng/10a01.pdf [accessed January 13, 2015].

UNFCCC/CMP (2012b) Report of the Conference of the Parties serving as the meeting of the Parties to the Kyoto Protocol on its seventh session, held in Durban. Action taken by the Conference of the Parties serving as the meeting of the Parties to the Kyoto Protocol: Addendum 2. FCCC/KP/CMP/2011/10/Add.2. http://unfccc.int/resource/docs/2011/cmp7/eng/10a02.pdf [accessed January 13, 2015].

UNFCCC/CMP (2013a) Report of the Conference of the Parties serving as the meeting of the Parties to the Kyoto Protocol on its eighth session held in Doha. Action taken by the Conference of the Parties serving as the meeting of the Parties to the Kyoto Protocol: Addendum 1. FCCC/KP/CMP/2012/13/Add.1. http://unfccc.int/resource/docs/2012/cmp8/eng/13a01.pdf [accessed January 13, 2015].

UNFCCC/CMP (2013b) Report of the Conference of the Parties serving as the meeting of the Parties to the Kyoto Protocol on its eighth session. Action taken by the Conference of the Parties serving as the meeting of the Parties to the Kyoto Protocol: Addendum 2. FCCC/KP/CMP/2012/13/Add.2 http://unfccc.int/resource/docs/2012/cmp8/eng/13a02.pdf [accessed January 13, 2015].

UNFCCC/SBI (2013) Compilation of information on nationally appropriate mitigation actions to be implemented by developing country Parties. FCCC/SBI/2013/INF.12/Rev.2.   http://unfccc.int/resource/docs/2013/sbi/eng/inf12r02.pdf   [accessed December 16, 2014].

UNFCCC/CMP (2014) Report of the Conference of the Parties serving as the meeting of the Parties to the Kyoto Protocol on its ninth session, held in Warsaw. Action taken by the Conference of the Parties serving as the meeting of the Parties to the Kyoto Protocol: Addendum 1. FCCC/KP/CMP/2013/9/Add.1. http://unfccc.int/resource/docs/2013/cmp9/eng/09a01.pdf [accessed January 14, 2015].

UNFCCC/CMP (2015) Report of the Conference of the Parties serving as the meeting of the Parties to the Kyoto Protocol on its tenth session, held in Lima. Action taken by the Conference of the Parties serving as the meeting of the Parties to the Kyoto Protocol: Addendum 1. FCCC/KP/CMP/2014/9/Add.1. http://unfccc.int/resource/docs/2014/cmp10/eng/09a01.pdf [accessed February 20, 2015].

UNGA (1988) *Protection of Global Climate Change for Present and Future Generations of Mankind.* Resolution 43/53. December 6, 1988.

UNGA (1989a) *Possible Adverse Effects of Sea-Level Rise on Islands and Coastal Areas, Particularly Low-Lying Coastal States.* Resolution 44/206, December 22, 1989.

UNGA (1989b) *Protection of the Global Climate for Present and Future Generations of Mankind.* Resolution 44/207, December 22, 1989.

UNGA (1990) *Protection of the Global Climate for Present and Future Generations of Mankind.* Resolution 45/212, December 21, 1990.

Vidal, J. and Harvey, F. (2013) Green Groups walk out of UN Climate talks. *The Guardian,* 21 November 2013.

Vitelli, A. and Nicola, S. (2013) "UN global warming talks blocked by Russia set back six months. *Bloomberg News,* June 14, 2013. http://www.bloomberg.com/news/articles/2013-06-14/climate-talks-failure-risks-2015-deadline-on-emissions-pact-1- [accessed May 11, 2015].

WMO (1989) *The Changing Atmosphere. Implications for Global Security.* WMO-No. 710, Geneva: WMO.

WMO, World Climate Conferences. http://www.wmo.int/pages/themes/climate/international_wcc.php [accessed April 21, 2014].

World Bank (2012) *Turn down the Heat.* Washington, DC: World Bank.

3

Where's the "Energy" in Key UN Global Outcomes on Sustainable Development?

Examining the Record from UNCHE 1972 to Rio+20 2012 for References to the Nexus Between Climate Change and Energy Access for the Poor

There is no way to prove that any of this will happen until it actually occurs. The key question is: How much certainty should governments require before agreeing to take action? If they wait until significant climate change is demonstrated, it may be too late for any countermeasures to be effective against the inertia by then stored in this massive global system. The very long time lags involved in negotiating international agreement on complex issues involving all nations have led some experts to conclude that it is already late. Given the complexities and uncertainties surrounding the issue, it is urgent that the process start now.

UN (1987), Report on the World Commission on Environment and Development (Chap. 7, para. 23)

Energy and Global Climate Change: Bridging the Sustainable Development Divide,
First Edition. Anilla Cherian.
© 2015 John Wiley & Sons, Ltd. Published 2015 by John Wiley & Sons, Ltd.

3.1   Setting the stage: Why linkages between energy access for the poor and climate change matter for the UN's quest for sustainable development

Energy sustains human life in all parts of the globe. Demand for energy services to meet socio-economic development and improve human well-being is increasing as human populations grow. Cooking, heating, lighting, transport, communication, and pumping of water are just a few of the crucial energy services that enable socio-economic development around the world. Conversely, lack of access to energy services and the heavy reliance on inefficient energy services – particularly for cooking, heating, lighting, and obtaining clean water – impacts negatively on human well-being and consigns millions of people to lives plagued by ill-health.

Energy (sources, services, systems, and technologies) fuels growth at all levels, and that is why energy security and patterns of energy consumption and production are crucial elements of overall national development strategies in all countries. But it is also equally clear that the lack of access to cost-effective and reliable energy sources, services, systems, and technologies impinges on socio-economic growth within and across communities and countries. Modern sources of energy, which include natural gas, liquid petroleum gas, diesel and biofuels such as biodiesel and bioethanol, combust more efficiently, whilst the combustion of traditional biomass is both inefficient and engenders ill-health by contributing high levels of indoor air pollution amongst poor households that are dependent on biomass for meeting their basic energy needs.

Energy drives human development and anthropogenic global climate change. The energy sector is by far the largest source of greenhouse gases (GHGs), accounting for more than two-thirds of the total in 2010 – around 90% of energy-related GHG emissions are CO_2 and around 9% are methane (CH_4) (International Energy Agency [IEA], 2013). It is clear that countries around the world, no matter their population or the size of their economy, are interested in ensuring access to reliable and secure energy supplies that are cost-effective and preferably non-polluting to meet with their energy needs. But the serious development challenges of ensuring access to cost-effective, safe, and reliable energy services for all citizens are made that much more difficult for developing countries, particularly when combined with pressing resource (financial and capacity) constraints and the growing need to address energy sector-related climate change mitigation.

At the intergovernmental level, the vital role of energy in driving human development was explicitly recognized as early as the historic 1972 Stockholm Conference and the 1983 World Commission on Environment and Development. Goldemberg et al. (1985, 1988) provided an early understanding and recognition of the links between energy and a host of other development concerns including income poverty, malnutrition, and ill-health. Evidence underscoring these link-ages between energy and poverty, gender and health, in terms of the impacts of household energy use on indoor air pollution and the health of women and young children, was raised early by the World Health Organization (WHO, 2000). The role of access to energy in poverty reduction and socio-economic development has been well documented in the past (Leach, 1992; Sokona *et al.*, 2004; Modi *et al.*, 2005) and the idea that energy poverty contributes to hunger and indoor air pollution for the poor, who use more polluting and less effective energy sources, has been discussed by Sovacool (2012) and others.

The importance of increasing access to modern energy services for the poor is not a newly identified development challenge, as demonstrated in Chapter 1 and as will be discussed later in this chapter. Close to 20 years ago, Flavin and Lenssen (1994) argued that the risk of catastrophic climate change caused by growing carbon emissions from energy would be further challenged by the burgeoning energy needs of developing countries, and to meet these needs and also those of the global environment, incremental change in energy systems would clearly not be enough. For more than a decade, energy issues have been recognized to play a crucial role impacting on all three pillars – social, economic and environmental – of sustainable development, which is a concept that continues to be discussed and has great resonance at the intergovernmental level (Johansson and Goldemberg, 2002; Munasinghe, 2002). In other words, energy has a direct relationship with all three pillars of sustainable development: by fueling macroeconomic growth (economic); as a prerequisite for human well-being in terms of allowing for basic needs to be met (social); and as an environmental stressor causing pollution or contributing to GHG emissions, or, as in the case of some forms of renewable energy, lessening ecological stress (environmental). To be clear, dramatic variations in patterns of energy usage and access to modern, clean energy services have also been recognized as contributing to growing socio-economic inequities within and across countries. Noting that per-capita use of primary energy in developed countries was six times larger in developed countries than in developing countries more than a decade ago, a 2002 United Nations Development Programme (UNDP) report focusing on addressing the challenge of energy for sustainable development noted that extreme poverty accompanied by ill-health was worsened by the highly inefficient use of traditional biomass and poor energy services that are widespread in the developing world (Johansson and Goldemberg, 2002).

Arguably all countries are interested in ensuring that their energy needs are met in a cost-effective, reliable, and non-polluting manner, and that the energy choices and services offered are affordable to their consumers, secure from price and supply disruptions and safe in terms of consumption and use. In a globally interdependent world, no country is immune to or isolated from global fluctuations in energy prices and supplies, and no country is an energy "island" and, accordingly, policymakers seeking to advance the triumvirate of "energy security, economic and environmental objectives are facing increasingly complex – and sometimes contradictory – choices" (World Energy Outlook [WEO], 2012, p. 24). While this is true for policymakers in all countries, from the perspective of this chapter, it is important to highlight that policymakers from developing countries, particularly the least developed countries (LDCs) amongst them, are far more constrained in their abilities to make these complex policy choices simply because their countries are more vulnerable than others to the vagaries of the global energy market, and have greater capacity constraints and financing limitations in terms of addressing energy needs, including those for poverty reduction and sustainable development. For developing countries in general, and LDCs in particular, lack of access to cost-effective and reliable energy sources, services, systems, and technologies impacts adversely on socio-economic growth, and worsens development challenges, including poverty and health concerns, as well as gender inequality at all levels within and across countries.

What this chapter argues is that the nexus between increasing energy access for the poor and mitigating climate change is important for the broader UN-led global community embarked on finding an integrated post-2015 development agenda. The development challenges associated with the lack of energy access for the poor have been highlighted in Chapter 1. This chapter focuses on examining the references to, and linkages between, energy access for the poor and climate change within the context of key UN-led global conferences on environment and sustainable development over the course of 40 years, starting with the 1972 Stockholm Conference and going through to the 2012 Rio+20 Summit. How exactly has energy and, more specifically, the issue of energy access for the poor been addressed in key UN globally agreed outcomes over the course of four decades?

This chapter will provide evidence that the challenge of increasing access to modern energy services has been well recognized and studied within the UN context, but evidence derived directly from key globally agreed outcomes indicates that this issue has not been given much prominence. The aim of this chapter is not to assign responsibility for why energy access for the poor has not been adequately referenced in agreed global outcomes, or to provide a descriptive analysis of the UN sustainable development negotiations. Instead, the chapter has a more circumscribed scope, which is to examine the actual textual record of agreed outcomes to see exactly how energy issues and, in particular, energy access for the poor have been addressed. In so doing, the chapter hopes to demonstrate that despite a broad global consensus that the lack of access to energy services impacts negatively on the lives of the poor, the issue of energy access has not been given due policy and programmatic prominence in terms of UN-agreed outcomes on sustainable development over a long period of time. The absence of concrete references linking the energy–poverty–climate change nexus in past and ongoing UN-led discussion on sustainable development is arguably a waste of valuable resources and time. The existence of two distinct and separate negotiating tracks on energy for sustainable development and climate change that simply do not cross-reference results in outcomes are silos that do not lend themselves to the overall UN quest for a shared and integrated post-2015 development agenda.

This chapter and the previous chapter are intended to provide evidence for the argument that concrete references and stated linkages between energy access for the poor and climate change are largely missing in the two different tracks of intergovernmental negotiations related to climate change and sustainable development. Chapter 2 focused exclusively on examining the agreed outcomes that have emanated from the separately convened UN-led negotiations on climate change that have emerged since the 1990s. This chapter focuses on the key UN global outcomes that have emerged as a consequence of intergovernmental negotiations on the environment and development and sustainable development over the course of 40 years. This chapter will examine the agreed global record of key UN-led negotiations on sustainable development to see how and whether linkages between energy access for the poor and climate change have been addressed.

Given the central role that the energy sector plays in both ongoing and future socio-economic development of all countries across the globe, it would appear at first glance that energy issues should play a central role in the intergovernmental negotiations on development in general, and sustainable development in particular. Chapter 2 examined the record of more than 20 years of global climate change negotiations to see what, if any, references were made in these outcomes to the linkages between energy and poverty reduction, but the focus of this chapter will be to examine the agreed global outcomes of a different set of high-level UN conferences focused on environment and development and sustainable development to see how and if energy access, poverty reduction, and climate change feature in them. The fundamental questions explored in this chapter are to examine the record of key UN global environment and sustainable development outcomes over time, to see how this issue of increasing energy access for the poor has been dealt with, and to examine whether and to what extent these agreed UN global outcomes reference and include specific linkages between energy and two of the most pressing global development challenges facing the world today – poverty reduction and global climate change.

In order to understand how energy issues have been dealt with in terms of UN-led global outcomes on sustainable development, it is important first to consider how energy as an issue has been, and is being, dealt within the UN as an organizational system of agencies. Accordingly, a brief overview of the UN in terms of its current organizational rubric on energy is provided in the following section.

The chapter then provides a detailed examination of key UN global conferences over the course of 40 years, where energy, poverty reduction, and climate change have been considered. It also examines energy for sustainable development negotiations convened by the UN's Commission on Sustainable Development (CSD). More recent UN-sponsored initiatives, such as the Sustainable Energy for All (SE4All) and the Global Sustainability Panel (GSP), are also considered. Finally, the chapter concludes by providing an overview of its findings and discusses energy access for the poor and climate change-related targets in the 2014 global negotiations on sustainable development goals (SDGs).

3.2   The locus of "energy" within the UN context: Framing the issue of energy access for the poor at the global level

At the intergovernmental level within the context of the UN, "energy" is not listed as a stand-alone sectoral or thematic issue, but instead is listed as one amongst a wide range of development issues. This is important to highlight from the outset, because the organizational context by which energy is referenced at the UN does have an influence on the scope for intergovernmental discussion and action on energy for sustainable development. Energy for sustainable development is the broad category under which energy issues have been and continue to be discussed within the UN context. From 2000 onwards, UN-led global discussions have focused on the linkages between energy and sustainable development through the forum of the UN's CSD (UN, 1999), and also within the context of the 2002 World Summit on Sustainable Development held in Johannesburg and the 2012 UN Conference on Sustainable Development, also known as the Rio-20 Summit, held in Rio de Janeiro.

From an organizational perspective of the UN, as per the official UN website (UN, 2013), the issue of "development" is listed as one of five key issues that the UN focuses on, with the other four issues listed as peace and security, human rights, humanitarian issues, and international law. Within the broad category of "development", the UN currently lists 11 thematic areas, including advancement of women; countries in special situations; governance and institutional-building; international trade; macroeconomics and finance; population; science, technology and productive sectors; social development; statistics; sustainable development, human settlements and energy; and topics A–Z (UN system programs).

It is worth noting that the range of thematic areas covered under the rubric of "development" is wide and far-reaching and that there is no immediate rationale as to why one thematic area appears to be a catch-all categorization more than a specific thematic issue because it references three very distinct topics, namely "sustainable development, human settlements and energy", which are each very different in terms of substantive, as well as programmatic, scope and content. In fact, this particular thematic area is especially difficult to understand from an informational perspective, precisely because sustainable development is a more globally significant and referenced topic that cuts across most of the other thematic issues. The idea that energy could be lumped along with human settlements as part of a triad of issues is equally puzzling, because there does not appear to be a unifying conceptual understanding of why these three particular issues should be coalesced under one category.

As one of a triad of issues listed under a broad thematic category, it is also not immediately clear how energy issues are dealt with from a policy and implementation perspective within the wide context of the UN family of agencies. In terms of the UN organizational schematic, energy issues are not listed as falling under the purview of any one particular global body. From the perspective of the UN, policy and programmatic inputs related to a wide variety of energy for sustainable development topics have emanated from UNDP and UN Environment Programme (UNEP), while the negotiating forum for this topic was managed by UN Department of Economic and Social Affairs (UNDESA).

The UN bodies listed as the global bodies charged with the thematic area of "sustainable development, human settlements and energy" are the CSD, the UN Forum on Forests, the Governing Council of the UNEP, and the Executive Board of the UNDP/UN Fund for Population Affairs. In addition, the schematic lists nine global programmes conducted by various UN agencies and institutions, which may be clustered together as follows:

- sustainable development and sustainable forest management (UNDESA);
- environment and energy (UNDP);
- ecosystem management and harmful substances and hazardous waste (both under management by UNEP);
- water and sanitation, sustainable cities, and urban management (all managed by the UN Human Settlements Programme).

Additionally there are five regional programmes listed as being conducted by the regionally derived economic commissions, with only the Economic Commission for Europe having a program explicitly entitled "sustainable energy" (UN website, *Sustainable Development, Human Settlements And Energy*). Based on this organizational breakdown, it would appear at first glance that only the UNDP has a specific global program on environment and energy, but the reality is that energy issues are handled by different UN agencies, and from the perspective of the intergovernmental negotiations on energy for sustainable development, it has been the CSD managed by UNDESA that has served as the principal forum for global discussions on energy for sustainable development.

Recognizing the need for internal UN coordination, UN-Energy was established in 2004 as a response to the 2002 World Summit on Sustainable

Development and was specifically mandated to promote inter-agency collaboration in the field of energy, to support countries in their transition to sustainable energy, and to increase engagement with non-UN multistake-holders (http://www.un-energy.org/). Arguably, the aim of UN-Energy (where Secretariat services are provided by UNDESA) is to promote system-wide collaboration in the area of energy, as there is no single entity in the United Nations system that has primary responsibility for energy issues. Its stated role is to increase the sharing of information, encourage and facilitate joint programming, and develop action-oriented approaches to coordination. UN-Energy's work is organized around three thematic clusters, each led by two UN organizations:

- energy access: led by UNDESA and UNDP, in partnership with the World Bank;
- renewable energy: led by the Food and Agriculture Organization (FAO) and UNEP, with support of the UN Educational, Scientific and Cultural Organization (UNESCO);
- energy efficiency: led by the United Nations Industrial Development Organization (UNIDO) and the International Atomic Energy Agency (IAEA).

Interestingly, the UN lists a separate knowledge based tool for climate change, which is devised more as information-based portal, and not as an inter-agency mechanism for coordination- namely, the UN Climate Change Gateway. The informational and organizational separation and distinction, between energy and climate change issues in both knowledge based and programmatic terms is important to highlight and keep in mind particularly in the following sections which discuss intergovernmental energy for sustainable development negotiations.

Finally, it should also be pointed out that within the UN context, the intergovernmental negotiations on energy for sustainable development have been conducted under the auspices of the CSD. The CSD was the organizational forum envisioned by Agenda 21, which was the program of action adopted by the 1992 UN Conference on Environment and Development (UNCED). Agenda 21 called for the creation of the CSD to ensure effective follow-up of the UNCED; to enhance international cooperation and rationalize intergovernmental decision-making capacity; and to examine progress in Agenda 21 implementation at the local, national, regional, and international levels.

As a subsidiary body of the UN Economic and Social Council (UNECOSOC), the CSD comprised 53 rotating members, although all UN member countries were allowed to participate in CSD meetings and discussions. The election of members to the CSD was based on geographical allocation requirements: 13 seats from Africa, 11 seats from Asia, six seats from eastern Europe, 10 seats from Latin America and the Caribbean, and 13 seats from western Europe and

North America. Other states, organizations in the UN system, and accredited intergovernmental organizations and non-governmental organizations (NGOs) can attend Commission sessions as observers. The Division for Sustainable Development in the UNDESA served as the CSD Secretariat. The CSD, which was formally established in 1992 by UN General Assembly Resolution 47/191, held its first substantive session in June 1993 and met once annually between 1993 and September 2013, when it was replaced by the High Level Political Forum (HLPF). Like its predecessor, the CSD, which was outcome of Agenda 21, the HLPF has been initiated as an institutional outcome of the Rio+20 Outcome document entitled *The Future We Want*. The energy for sustainable development negotiations conducted within the CSD, and the anticipated role of energy within the HLPF in establishing the post-2015 sustainable development agenda, is discussed later on in this chapter.

In order to frame the discussion on energy access at the UN, it is important to note that the term "energy access" has been defined differently based on various conceptualizations of what the term "access" means, and this poses a challenge in terms of data collection and implementation of energy access goals at the country level. The lack of a clear and succinct definition of energy access over time has resulted in some ambiguity in terms of how individual developing countries understand and respond to energy access policies and implementation, and also in terms of temporal data comparison both across countries and within countries. This definitional challenge as to what exactly is meant by energy access for all – or universal energy access – continues to impact on the way in which energy access for the poor is understood and implemented at the global and national level. Srivastava *et al.* (2012) have argued that although the links between lack of energy access and poverty were known, renewed attention on energy occurred only when energy was stated to be essential to meeting the Millennium Development Goals (MDGs) and it was referenced as the missing MDG. Some of the most comprehensive studies on energy access in developing countries were done by the largest development agency of the UN, the UNDP (2007a–c). The challenge of defining energy access and agreeing on a set of related goals and targets that would allow for increased energy access by the poor has also been highlighted in special issue of *Energy Policy* entitled "Universal access to energy: Getting the framework right" (Srivastava and Sokona, 2012), with contributions from energy experts principally from Africa and South Asia, the two regions of the world with the largest concentrations of the energy poor.

A landmark 2009 UNDP/WHO report on the energy access situation in developing countries highlighted the difficulties of defining the scope and complexity of energy access, but did not provide any succinct definition, and instead listed a series of indicators used to measure energy access, namely:

- access to electricity which is measured as the percent of people who have a household electricity connection;

- access to modern fuels measured as the percent of people who use electricity, liquid fuels, or gas as their primary fuel to satisfy cooking needs;
- access to mechanical power measured as the percent of people who use mechanical power for productive, non-industrial applications, such as water pumping and small-scale agro-processing.

In addition, to better contextualize energy access in relation to cooking needs in developing countries, the report also references data collection related to access to improved cooking stoves, measured as the percent of people relying on solid fuels – traditional biomass and coal – and access to cooking fuels measured as the percent of people who use different types of cooking fuels as their primary cooking fuels (UNDP/WHO, 2009). But the real significance of this report lies in its list of findings (see Figure 3.1), which provide a clear set of linkages between energy access and climate change on the poor.

The report's findings, that targets for increasing energy access and improved data on energy access are both urgently needed, as well its explicit recognition of the global warming impacts of poorly combusted solid fuels, are particularly significant for the purpose of this chapter and the overall argument of this book, but interestingly, as the chapter will evidence, this issue would not be reflected in the 2012 Rio+20 Summit despite a UN-led initiative to advocate for universal energy access for all. The scope and scale of the list of findings contained in the Box 3.1 provide the rationale for proposing urgent and effective action expressly focused on promoting sustainable energy services for the poor in Chapter 5 of this book.

The linkages between energy access and poverty reduction were made even more explicit in a special report entitled *Energy Poverty: How to Make Modern Energy Access Universal*, which noted that it was "both shameful and unacceptable that billions of people lack access to the most basic energy services with the situation most like to deteriorate in the next 20 years" and reiterated the fact that 1.4 billion people, close to 20% of the global population, lack access to electricity, and that 2.7 billion people, or close to 40% of the global population, rely on the traditional use of biomass and its attendant causes of household ill-health and pollution (IEA/UNDP/UNIDO, 2010). In both this report and the 2009 UNDP/WHO report, there is an explicit recognition that sub-Saharan Africa is the region where energy access is most urgently needed and that heavy reliance on traditional and inefficient biomass sources engenders ill-health and increased mortality, particularly amongst poor women and children. Making the case for urgent action by the global community, the report provides an assessment of two key indicators of energy poverty at the household level: the lack of access to electricity and the reliance on the traditional use of biomass for cooking. Its key findings are summarized in Box 3.1, but it is worth highlighting that the last two findings are related to investments costs needed for energy access and that the resulting increase in primary energy demand and CO_2 emissions is estimated to be modest.

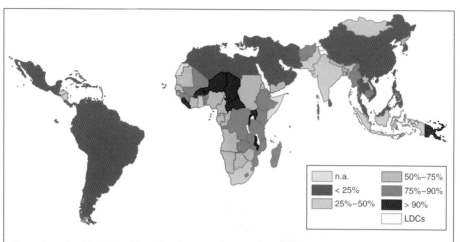

Share of people without electricity access in developing countries, 2008

❖ 79% of people in least developed countries (LDCs) and 74% of people in sub-Saharan Africa lack access to electricity.

❖ 56% of people living in developing countries rely primarily on solid fuels – coal and traditional biomass– for their cooking needs and only 27% of those who rely on solid fuels are estimated to use improved cook stoves.

❖ Worldwide almost two million deaths annually from pneumonia, chronic lung disease, and lung cancer are associated with exposure to indoor air pollution resulting from cooking with biomass and coal, and 99% of them occur in developing countries. Close to half (45%) of the global population relies on solid fuels and 44% of those who die from the resultant exposure to indoor air pollution are children, while women account for 60% of all adult deaths.

❖ Emissions from burning solid fuels in open fires and traditional stoves also have significant global warming effects, due to incomplete combustion fuel carbon.

❖ Developing countries are far behind in expanding access to modern energy, whether to meet nationally set energy access targets or facilitate achievement of the Millennium Development Goals. About half of developing countries have targets for improving access to electricity, but few countries have targets for modern fuels, improved cook stoves or mechanical power, which are all crucial for meeting the basic needs of the poor.

❖ Continued efforts are required to improve the quantity and quality of statistical information related to energy access. Particularly noteworthy is the lack of data available on improved cook stoves, which 73 of 140 countries lack data on, and mechanical power, for which 137 of 140 countries lack data.

❖ Greater broad-based efforts are needed to expand access to modern energy services to those who lack access, especially to heat for cooking and to mechanical power in rural and remote areas.

Figure 3.1   Key findings: energy access in developing countries (source: UNDP/WHO, 2009).

Before beginning an examination of how energy issues and energy access have been reflected in key UN global outcomes related to sustainable development, it is useful to note that the issue of energy access has been getting increasing global prominence since 2010, as a result of the UN Secretary-General's establishment of an Ad Hoc Advisory Group on Energy and Climate Change (AGECC), comprising UN agencies and stakeholders from civil society (research institutions, non-governmental organizations, and the private sector). Additional global attention was brought to bear as a result of an influential special report on energy access, entitled *Energy for All: Financing Access for the Poor*, prepared by the IEA (2011), which was released

Box 3.1   *Key challenges: making modern energy access universal*

- Greatest challenge lies in Sub-Saharan Africa where the electrification rate is 31% and the number of people relying on the traditional use of biomass is 80%.
- Today, there are 1.4 billion people around the world that lack access to electricity, some 85% of them in rural areas. Without additional dedicated policies, by 2030 the number of people drops, but only to 1.2 billion. Some 15% of the world's population still lack access, the majority of them living in Sub-Saharan Africa.
- The number of people relying on the traditional use of biomass is projected to rise from 2.7 billion today to 2.8 billion in 2030.
- It is estimated that household air pollution from the use of biomass in inefficient stoves would lead to over 1.5 million premature deaths per year, over 4,000 per day, in 2030, greater than estimates for premature deaths from malaria, tuberculosis or HIV/Aids.
- Addressing these inequities depends upon international recognition that the projected situation is intolerable, a commitment to effect the necessary change, and setting targets and indicators to monitor progress. A new monitoring tool – the Energy Development Index (EDI) – that ranks developing countries in their progress towards modern energy access is provided.
- The UN Millennium Development Goal of eradicating extreme poverty by 2015 will not be achieved unless substantial progress is made on improving energy access. To meet the goal by 2015, an additional 395 million people need to be provided with electricity and an additional one billion with access to clean cooking facilities. This will require annual investment in 2010–2015 of $41 billion, or only 0.06% of global GDP.
- To meet the more ambitious target of achieving universal access to modern energy services by 2030, additional investment of $756 billion, or $36 billion per year, is required. The resulting increase in primary energy demand and CO_2 emissions would be modest. In 2030, global electricity generation would be 2.9% higher, oil demand would have risen less than 1%, and CO_2 emissions would be 0.8% higher.

Source: IEA/UNDP/UNIDO (2010, p. 7).

at a global conference on the topic convened by the Government of Norway. What is striking is that this 2011 report points out that "there is no universally-agreed and universally-adopted definition of modern energy access" and goes on to define modern energy access as "a household having reliable and affordable access to clean cooking facilities, a first connection to electricity and then an increasing level of electricity consumption over time to reach the regional average", based on an assumption of an average of five people per household (IEA, 2011, p. 12) By defining energy access in terms of access at the household level, the IEA report recognized that factors such as electricity usage in public buildings, which are important loci for socio-economic development such as schools and hospitals, for example, are excluded. In an effort to further extend its definition, the IEA noted that its definition involves a specified minimum level of electricity per household, varied for rural (250 kilowatt-hours [kWh]/year) and urban (500 kWh/year) areas (IEA, 2011). According to the IEA, in rural households, this minimum level of consumption could support the use of the following: a floor fan, a mobile telephone, and two compact fluorescent light bulbs for about 5 hours/day, whereas an urban household would also include a small, efficient refrigerator, a second mobile phone and another household appliance such as a small television or computer (IEA, 2011).

The IEA report confirmed the reality that millions of poor households were far removed from these idealized definitional threshold levels, and it estimated that 1.3 billion people have no access to electricity, and 2.7 billion people lack access to clean cooking facilities. The bulk of these populations of the energy-poor are concentrated in two particular regions of the world: more than 95% are either in sub-Saharan Africa or developing Asia, and 84% are in rural areas. Furthermore, the report noted that of the 1.9 billion who rely on traditional biomass for cooking, around 840 million live in India and more than 100 million live in Bangladesh, Indonesia, and Pakistan, whilst over 100 million people in Nigeria have no access to clean cooking facilities (IEA, 2011). Given the important role played by the IEA at the global level in both energy for sustainable development and global climate change negotiations, what these estimates indicate is that future policy and programmatic action on energy access needs to take stock of the regional representation of the energy-poor. In terms of the discussion in Chapters 4 and 5 of this book, the reliance of the poor on traditional biomass and the attendant health and environmental problems caused by indoor air pollution and black carbon emissions mean that future synergies on energy and climate action need to be focused in terms of increasing access to sustainable energy services in sub-Saharan Africa and South Asia, where the needs and impacts are the greatest.

Regardless of the definitional challenges, two crucial indicators of energy poverty at the household level are the lack of access to modern, clean energy sources, and the reliance on the traditional use of biomass for household cooking, both of which seriously constrain the productive opportunities of the poor and have grave negative impacts on their health and well-being. In reviewing the energy access situation in developing countries, WHO/UNDP (2009) found that access to energy is much lower in poorer developing countries and in certain regions, with "79 percent of people lacking access to electricity in LDCs, and 74 percent in sub-Saharan Africa, compared to 28 percent in developing countries as a whole". According to WHO/UNDP (2009), 45% or "almost half of the global population still relies on solid fuels for household use", which is associated with "almost 2 million deaths annually from pneumonia, chronic lung disease and lung cancer" caused by exposure to indoor air pollution resulting from biomass and coal usage, but in "LDCs and sub-Saharan Africa, more than 50 percent of all deaths from these three diseases can be attributed to solid fuel use, compared with 38 percent in developing countries overall" (p. 1).

But more recently, projections from the WEO suggest that the problem will not only persist, but in fact deepen in the longer term without an international recognition and commitment to effect change (WEO, 2011). Recognizing that metrics matter and that data collection is essential to track progress and for comparative purposes, Nussbaumer *et al.* (2012) investigated the concept of energy poverty and ladders, and proposed a new composite index to measure energy poverty – the Multidimensional Energy Poverty Index (MEPI). Unlike other indices that focus on more subjective assessments of energy access, they argue that their index captures both the incidence and intensity of energy poverty and can be used as a policymaking tool and offer initial application results for several African countries.

Within the context of the UN, energy access issues have largely been addressed within the broader framework of energy for sustainable development, but there has not been any detailed examination of the record of key landmark globally agreed outcomes to see exactly how the issue of energy access has been referenced within these outcomes. The following section provides a detailed examination of the energy–poverty–climate change nexus by carefully reviewing all the major UN conferences on environment and sustainable development over the course of four decades. The aim is to provide a careful accounting of references and stated linkages between energy, poverty reduction, and climate change in key UN global conferences on environment and sustainable development over a 40-year period, starting from the 1972 United Nations Conference on the Human Environment and ending with the 2012 Rio+20 Summit and the UN follow-up processes related to defining a set of SDG) and a framework for a post-2015 development agenda.

3.3 Examining the energy–poverty–climate change nexus in key UN global conferences on environment and sustainable development from 1972 to 2000

The UN serves as the sole forum for intergovernmental negotiations on sustainable development, which is where issues such as energy access, poverty reduction and climate change have been primarily considered. In order to understand how energy for sustainable development, and, in particular, energy access for the poor, has been referenced at the UN, a series of historic UN global conferences and summits on environment and sustainable development from 1972 to 2012 will be examined. The agreed global outcomes of this series of historic and global UN conferences serve as the primary source material for understanding whether and how the energy access for the poor and climate change nexus has been referenced. The agreed global outcomes of key UN conferences from the 1972 Stockholm Conference to the 2012 Rio+20 Summit serve, therefore, as significant benchmarks of what was agreed to by the global community in a particular time frame. These agreed global outputs have never before been specifically examined in terms of a 40-year chronological record to see what, if any, are the references to, and linkages between, energy access for the poor and climate change.

Before examining the record of these key UN outcomes on sustainable development, it is important to reflect briefly on the institutionalization of the UN-led process for convening massive global conferences or world summits and the role of developing countries' participation in UN sustainable development-related processes.

Klein (2005, pp. 3–4) pointed out that from the 1992 Earth Summit to the World Summit on the Information Society of 2003, the UN hosted almost one summit per year for 11 years, and stated that all such world summits use a broadly similar "form", which defines the processes of participation and engagement at the summit while the "content" or sets of issues and the results emanating from various summits are often different. Convening world summits or global conferences has become the globally acceptable norm within the context of the UN development agenda. These conferences have often resulted in number of spin-off issues, which in turn have sponsored a separate trajectory of periodic global meetings and conferences. Some of these spin-off issues have resulted in the implementation of substantial and new multilateral environmental agreements as in the case of 1992 Rio Summit and the subsequent United Nations Framework Convention on Climate Change (UNFCCC) and the UN Convention on Biological Diversity (UNCBD; Cherian, 2012).

The reality is that in some years there has been more than one world summit or special global conference (refer to the UN website, www.un.org):

- 2000 – Millennium Summit, World Education Forum, Social Summit +5, Special Session of the General Assembly for the 5 Year Review of the Beijing and Platform for Action;
- 2002 – International Conference on Financing for Development, World Summit on Sustainable Development, Second World Assembly on Ageing, Special Session of the General Assembly on Children and World Food Summit: Five Years Later;
- 2005 – World Summit, 10 Year Review of the Copenhagen Declaration and Programme for Action, 10 Year Review of the Beijing Declaration and Platform for Action, Phase Two of the World Summit on the Information Society;
- 2008 – Follow-up International Conference on Financing for Development to Review the Implementation of the Monterrey Consensus and High Level Event on the Millennium Development Goals.

Hosting and convening these conferences are costly and time-consuming but there are no globally agreed benchmarks for evaluating summit outcomes over an extended period of time, because the scope of the issues and the extent to which legally binding agreements are defined across different multi-lateral thematic concerns vary so widely.

Another important point to note is that the burgeoning growth in UN-related global fora on a variety of sustainable development issues has not resulted in lessening of developing countries' engagement in these fora despite existing resources and capacity constraints. Negotiators from developing countries have been actively engaged with the global debate on sustainable development as evidenced by their early participation in Stockholm 1972 which has escalated with subsequent UN summits in 1992, 2002, and, more recently, in 2012. Najam (2005) argued that developing countries' involvement can be characterized in the pre-Stockholm era as a "politics of contestation", in the Stockholm-to-Rio period as a " period of reluctant participation", and in the Rio to Johannesburg period as "the emergence of more meaningful, but still hesitant, engagement" (p. 303). But rather than characterizing developing countries' involvement as moving along some evolutionary path, it is possible to argue that leadership roles in the negotiations have been taken by developing country negotiators early in the process. The idea that somehow long-standing divisions between developed and developing country negotiators over assigning responsibility for addressing the global climate change problem can be relegated to a pre-Rio 1992 ear belies the 2009 impasse in the Copenhagen Conference of the Parties (COP).

The following section does not seek to provide a historical accounting or analytic narrative of the negotiating processes associated with each of the major UN outcomes, ranging from the 1972 United Nations Conference on Human Environment (UNCHE) to the 2012 Rio+20 Summit. The aim is not to provide a description of the negotiations, or to highlight what a particular set of actors (state or non-state) and/or institutional norms and practices may or may not have contributed or done in a particular globally agreed outcome. There is already available a rich analysis and review of most of the major UN global conferences available. For instance, Haas *et al.* (1992) have evaluated the 1992 UNCED in terms of its policy contributions in institutionalizing the broader principle of sustainable development. Jordan and Voisey (1998) have focused on the specific processes and outcomes of the 1992 Rio Summit. Chasek and Wagner (2012) have provided an in-depth historical analysis of 20 years of key UN multilateral negotiations processes and outcomes since the 1992 Rio Summit.

3.3.1 Examining the energy–poverty–climate change nexus in the 1972 UNCHE

The convening of the 1972 UNCHE in Stockholm – the UN's first major global conference on environment and development – was significant in many different ways. It set the precedent for an agreed UN environment/sustainable development outcome that included a political declaration followed by a series of principles relating to either one or a series of global environmental challenges. With its motto of, "Only one Earth", the UNCHE Declaration contained a common set of 26 principles aimed at guiding the preservation and enhancement of the human environment. The two broad ideas reflected in the 1972 Stockholm Declaration that were considered novel at their time, and are particularly noteworthy for their continued resonance as definitive features of global multilateral environmental agreements (MEAs), are as follows (Cherian, 2012, p. 43):

a. The need to defend and improve the human environment for present and future generations, as well as,
b. The recognition of the growing emergence of new set of regional and global environmental problems that defy national boundaries, and by their global or regional scope and nature necessitate cooperative, intergovernmental responses.

In analyzing the Stockholm Conference, Sohn (1973, p. 423) noted that the UNCHE was, in many respects, one the most successful international conferences held at the time, although it is not possible to compare across different UN conferences given the growth in multilateralist environmental awareness amongst states and civil society institutions, and active civil society and state engagement from the 1990s onwards. From a developing country perspective, a key institutional outcome resulting from the UNCHE was the adoption of the UN General Assembly adoption of Resolution 2997 on December 15, 1972 creating the UNEP and its Governing Council composed of 58 nations elected for 4-year terms by the UN General Assembly, and assigning UNEP responsibility for assessing the state of the global environment and for serving as a focal point for environmental action and coordination within the UN system, and, more importantly, for locating its Secretariat in Nairobi, Kenya, which was the first and, to this date, the only UN hub/center in a developing country.

What is also remarkable about the UNCHE is that for the first time developing countries, including Brazil, India, and China, and civil society actors (non-governmental organizations) were involved in an open multilateral forum for discussions, including one in which Prime Minister Indira Gandhi of India stated that poverty was the main cause of environmental degradation (Strong, 1973; Caldwell, 1984). This idea that poverty reduction is crucial to sustainable development needs of developing countries continues to resonate as a central issue over 40 years later in the 2012 Rio+20 outcome, and in ongoing UN discussions on shaping the post-2015 development agenda.

The UNCHE also set a new threshold in terms of environmental governance for including a total of 109 recommendations encompassing a wide range of issues from organizational matters to germ plasm and water resources. It is worth noting that Recommendations 57–59 (see Box 3.2) focused on energy, and that these recommendations were quite specific in their scope and coverage, including the linkages between energy production and consumption and environmental effects.

Notably, the UNCHE's call for data and monitoring on the environmental effects and consequences of energy systems and energy use and production, including the monitoring of emissions of carbon dioxide and other gases, can be seen as the first intergovernmental guidelines for monitoring energy-related emissions of gases. For instance, Recommendation 57 can be seen as an early precursor to GHG emissions monitoring, in that it calls for monitoring of CO_2 etc., including releases from oil. Recommendation 58, on the other hand, represented a unique and early opportunity for the UN to serve as the equivalent of an informational clearing house or mechanism on energy, but this was never really followed through in the years and, interestingly, has recently surfaced in intergovernmental discussions on sustainable energy. In particular, the point that data on the environmental consequences of different energy systems should be provided through a variety of sources and fora, including national exchanges of information, can be viewed as the first UN global conference output where the need for data linkages between environment and energy were referenced – namely, the environmental aspects of energy systems. Finally, Recommendation 59, which discusses the need for a comprehensive study of energy sources, technologies, and trends, can be seen as an important influence and precursor for establishing what has become the annual World Energy Outlook – a key energy-related output of the Organisation for Economic Co-operation and Development (OECD)/IEA.

Box 3.2 UNCHE: Energy-related recommendations for action at the international level

Recommendation 57

It is recommended that the Secretary-General take steps to ensure proper collection, measurement and analysis of data relating to the environmental effects of energy use and production within appropriate monitoring systems.

a. The design and operation of such networks should include, in particular, monitoring the environmental levels resulting from emission of carbon dioxide, sulphur dioxide, oxidants, nitrogen oxides (NO.), heat and particulates, as well as those from releases of oil and radioactivity;
b. In each case the objective is to learn more about the relationships between such levels and the effects on weather, human health, plant and animal life, and amenity values.

Recommendation 58

It is recommended that the Secretary-General take steps to give special attention to providing a mechanism for the exchange of information on energy:

a. The rationalization and integration of resource management for energy will clearly require a solid understanding of the complexity of the problem and of the multiplicity of alternative solutions;
b. Access to the large body of existing information should be facilitated:

   i)   Data on the environmental consequences of different energy systems should be provided through an exchange of national experiences, studies, seminars, and other appropriate meetings;
   ii) A continually updated register of research involving both entire systems and each of its stages should be maintained.

Recommendation 59

It is recommended that the Secretary-General take steps to ensure that a comprehensive study be promptly undertaken with the aim of submitting a first report, at the latest in 1975, on available energy sources, new technology, and consumption trends, in order to assist in providing a basis for the most effective development of the world's energy resources, with due regard to the environmental effects of energy production and use: such a study to be carried out in collaboration with appropriate international bodies such as the International Atomic Energy Agency and the Organisation for Economic Cooperation and Development.

Source: UNEP website: http://www.unep.org/Documents.Multilingual/Default.asp? DocumentID=97&ArticleID=1506&l=en (accessed on October 14, 2014).

Missing from the UNCHE is any direct reference to the role of energy in promoting socio-economic development or reducing poverty in developing countries. The recommendations relating to energy are more focused on data and information gathering, and as there was no explicit linkage between energy and poverty reduction, there was accordingly no programmatic or policy references that called for or emphasized the need to ensure energy access for the poor or to address poverty reduction concerns. What is also useful to point out is that the UNCHE's recommendation to the UN Secretary-General in terms of data and monitoring systems related to environmental effects of energy use and production stated that the design and operation of networks should include "monitoring the environmental levels resulting from emission of carbon dioxide" amongst other gases. In other words, there was a clear statement regarding the environmental levels – thresholds – resulting from the emissions of gases related to energy use and production, even though the term "climate change" was not mentioned.

3.3.2   Examining the energy–poverty–climate change nexus in the WCED

The decade following the UNCHE witnessed the coinciding of the 1970s global energy crisis and growing global environmental awareness of the impacts of pollution. In December 1983, the UN General Assembly (UNGA) adopted a rather ambitious-sounding Resolution 38/61, entitled "Process of preparation of the Environmental Perspective to the Year 2000 and Beyond", which called for the creation of a Special Commission (UNGA, 1983). The Special Commission that resulted was the World Commission on Environment and Development (WCED), which is also referred to as the Brundtland Commission – named after its first chairperson, the Norwegian Prime Minister Gro Harlem Brundtland – but what is often neglected is that the Commission happened to include Mansour Khalid, the Foreign Minister of Sudan, as its Vice-Chair.

In 1987, the Brundtland Commission published its main report – *Our Common Future* – which would go on to have a very direct and influential impact on the scope of issues covered at the 1992 Earth Summit in Rio de Janeiro. In 1987, the UNGA adopted a procedural follow-up Resolution 42/186, entitled "Environmental Perspective to the Year 2000 and Beyond", which referenced the report of the WCED, but also included two clear linkages among poverty reduction, environmental degradation, and increasing energy access in paragraph 3.c, which stated, "Since mass poverty is often at the root of environmental degradation, its elimination and ensuring equitable access of people to environmental resources are essential for sustained environmental improvements", and in paragraph 4.c, which stated:

> The provision of sufficient energy at reasonable cost, notably by increasing access to energy substantially in the developing countries, to meet current and

expanding needs in ways which minimize environmental degradation and risks, conserve non-renewable sources of energy and realize the full potential of renewable sources of energy (UNGA, 1987a).

In Resolution 42/187 of December 11, 1987, which emphasized the need for a "new approach to economic growth, as an essential prerequisite for eradication of poverty and for enhancing the resource base on which present and future generations depend", the UNGA officially welcomed the report of the WCED entitled *Our Common Future*, but this resolution contained no references to energy whatsoever (UNGA, 1987b).

The WCED was one of the earliest globally organized initiatives that focused attention on global environmental issues, included an urgent appeal to reduce poverty, and also provided a clear and early warning of the dangers of global climate change. In its 1987 report, the concept of sustainable development was clearly defined for the first time. The report is often cited for its benchmark definition of sustainable development as "meeting the needs of the present without compromising the ability of future generations to meet their own needs", which is first contained in paragraph 27 of the overview section, and this definition is officially part of the first paragraph of Chapter 2 of WCED report, entitled *Towards Sustainable Development* (UN, 1987, Chap. 2, para. 1). The definition has stood the test of time and has been adopted as "a central guiding principle" of the UN, Governments and private institutions, organizations and enterprises by the UNGA and remains as such to this day.

This lasting definition of sustainable development also importantly contains two key and contingent concepts, namely (UN, 1987, para.1):

- the concept of "needs", in particular the essential needs of the world's poor, to which overriding priority should be given; and
- the idea of limitations imposed by the state of technology and social organization on the environment's ability to meet present and future needs.

The report's explanation of the concept of sustainable development begins with a categorical assertion that "sustainable development is intrinsically linked to the meeting of human needs" and that "a world in which poverty and inequity are endemic will always be prone to ecological and other crises. Sustainable development requires meeting the basic needs of all and extending to all the opportunity to satisfy their aspirations for a better life" (UN, 1987, para. 4).

Arguably this is the first clear evocation of the need for any global quest for sustainable development to address poverty reduction and ensure the meeting of the basic needs of all. This idea that sustainable development required the meeting of the basic needs of all and improving human well-being would have a direct impact not just on the scope of the 1992 Earth Summit in Rio, but also more importantly on the 2000 Millennium Declaration, whose express intent was to reduce poverty and improve maternal and child health, amongst other issues. This idea also continues to be referenced in most statements made by developing country delegates and leaders relating to sustainable development, including as recently as the 2013 UNGA, which considered the issue of setting the stage for the post-2015 sustainable development agenda.

The 1987 WCED report set a precedent in not viewing the concept of, and concern over, "environment" as separate from the concept of and concern over "development", arguing instead that these two issues were inseparable. From a historical perspective, this early definition of sustainable development accorded priority to the needs of the world's poor, and referenced limitations due to technology and human social capacities, both of which are negotiating references that continue to be raised more than 20 years later in most global negotiations associated with multilateral environmental agreements, ranging from the spread of desertification to biodiversity loss and global climate change.

In terms of energy, *Our Common Future* had a far greater and more explicit focus on energy than did the UNCHE. A review of the report reveals 438 references to energy but no explicit reference to the issue of energy access or access to energy. Meanwhile the term "poverty" is referenced 104 times in the report, but there is no direct mention of the terms "poverty reduction" or "poverty eradication". The issue of energy is quite clearly mentioned in the context of the WCED's focus on the linkages between ensuring essential needs for all and the broader global quest for sustainable development in a section entitled, "Meeting essential needs" – "Energy is another essential human need, one that cannot be universally met unless energy consumption patterns change" – and here the report is prescient in stating that, "by the turn of the century, 3 billion people may live in areas where wood is cut faster than it grows or where fuelwood is extremely scarce" (UN, 1987, Chap. 2, para. 46).

Unlike the UNCHE that preceded it, and the Agenda 21 that would come after it, the WCED contains a specific chapter – Chapter 7 – entitled "Energy: choices for environment and development". Chapters 7 included 8 subsections all together and was comprehensive in including "fossil fuels", "renewable energy", "nuclear energy", "energy efficiency", "energy conservation measures" and "wood fuels: the vanishing resource". There is no mistaking the importance of energy for sustainable development in terms of the content of this chapter, which begins by stating very simply that:

> Energy is necessary for daily survival. Future development crucially depends on its long-term availability in increasing quantities from sources that are dependable, safe, and environmentally sound. At present, no single source or mix of sources is at hand to meet this future need (UN, 1987, Chap. 7, para. 1).

This opening paragraph and the one that follows contain the first global acknowledgement that energy drives human development and that constraints (based on criteria ranging from safety to environmental concerns) in energy sources will impact future human development.

The highlighting and listing of climate change as one amongst four global risks (which also include nuclear reactor accidents and waste disposal) is what makes the report really stand apart from previously agreed global outputs. Box 3.3 provides an excerpt of key WCED findings on climate change, including the call for the creation of a convention. The report also notes four reservations associated with the high-energy future, including "the serious probability of climate change generated by the 'greenhouse effect' of gases emitted to the atmosphere, the most important of which is carbon dioxide (CO_2) produced from the combustion of fossil fuels; [and] urban-industrial air pollution caused by atmospheric pollutants from the combustion of fossil fuels". It also raises concern over the growing scarcity of fuel wood in developing countries (UN, 1987, Chap 7. para. 11). This issuance of an early warning signal touches on some of the challenges associated with increasing fossil fuel use and traditional biomass dependence.

In the sub-section dealing with fossil fuels, the report focuses on several key linkages between fossil fuel energy-related emissions and activities and "managing climatic change", and states up-front that "the risks of global warming make heavy future reliance upon fossil fuels problematic" (UN, 1987, Chap. 7, para. 19). The report's recommendations can be viewed as the first globally explicit recognition of the linkages between energy and climate change, and also as a precursor in terms of addressing the global climate change challenge. The report is succinct in stating that the burning of fossil fuels and the loss of negative cover, to a lesser extent, have increased the accumulation of CO_2 in the atmosphere. There is no mistaking its clarity in citing the 1985 expert meeting in Villach Austria which found:

> that concentration of CO_2 and other greenhouse gases in the atmosphere would be equivalent to a doubling of CO_2 from pre-industrial levels, possibly as early as the 2030s, and could lead to a rise in global mean temperatures 'greater than any in man's history (UN, 1987, Chap. 7, para. 21).

But perhaps the most remarkable part of the WCED in terms of its predictive salience is its early assertion of what would go on to become known as the precautionary principle. The report recognized early on both the complexity and the uncertainties associated, and laid out a four-track strategy, including improved monitoring and assessment of the evolving phenomena; increased research to improve knowledge about the origins, mechanisms, and effects of the phenomena; the development of internationally agreed policies for the reduction of the causative gases; and adoption of strategies needed to minimize damage and cope with the climate changes and rising sea level.

As the discussion on future UN conferences and global outputs related to sustainable development demonstrates, the WCED can be seen as the exception to the rule, in that it included explicit linkages between energy and climate change, and also contained the need to meet essential needs of all, including through

Box 3.3   Key Recommendations related to climate change: Excerpts from the Report of the World Commission on Environment and Development (UN, 1987, Chapter 7)

Implications of growth of GHGs emissions

- The pre-industrial concentration was about 280 parts of carbon dioxide per million parts of air by volume. This concentration reached 340 in 1980 and is expected to double to 560 between the middle and the end of the next century. Other gases also play an important role in this 'greenhouse effect', whereby solar radiation is trapped near the ground, warming the globe and changing the climate (para. 19).

- Combined concentration of CO_2 and other greenhouse gases in the atmosphere would be equivalent to a doubling of CO_2 from pre-industrial levels, possibly as early as the 2030s, and could lead to a rise in global mean temperatures 'greater than any in man's history'. Current modeling studies and 'experiments' show a rise in globally averaged surface temperatures, for an effective CO_2 doubling, of somewhere between 1.5°C and 4.5°C, With the warming becoming more pronounced at higher latitudes during winter than at the equator (para. 21).

- An important concern is that a global temperature rise of 1.5–4.5°C, with perhaps a two to three times greater warming at the poles, would lead to a sea level rise of 25–140 centimetres. A rise in the upper part of this range would inundate low-lying coastal cities and agricultural areas, and many countries could expect their economic, social, and political structures to be severely disrupted (para. 22).

Creating a convention

- Nations urgently need to formulate and agree upon management policies for all environmentally reactive chemicals released into the atmosphere by human activities, particularly those that can influence the radiation balance on earth. Governments should initiate discussions leading to a convention on this matter (para. 28).

- If a convention on chemical containment policies cannot be implemented rapidly, governments should develop contingency strategies and plans for adaptation to climatic change (para. 29).

Source: UN (1987).

energy services. The most straightforward explanation for this is that, although the WCED was adopted by the UNGA, the WCED itself was not the result of a typical UN negotiations process leading to a conference outcome document.

3.3.3   Examining the energy–poverty–climate change nexus in the 1992 Agenda 21

The historic UNCED was held in Rio de Janeiro, Brazil, from June 3 to 14, 1992. The UNCED, also referred to as the 1992 Rio Earth Summit, was convened after a period of 2 years of lengthy negotiations by the Preparatory Committees (PrepComs). At the UNCED, five major agreements on global environmental issues were signed. As a direct result of the UNCED and its outcomes, there have been two decades of intensive global negotiations on a range of environmental issues, from addressing climate change to biodiversity loss. UNCED outcomes, such as the UNFCCC and the UN Convention on Biological Diversity, had extensive intergovernmental negotiating processes prior to their being presented as formal treaties with legally binding provisions on parties who would sign on to them. The remaining three UNCED agreements were non-binding agreements and include the much-discussed Agenda 21, billed as a blueprint for sustainable development, the Rio Declaration which was a summary of key consensus-based principles relating to sustainable development, and the Statement on Forest Principles, which outlined the need for sustainable use of forest resources. Given the role of energy in driving human development and impacting on global climate change, energy issues should logically have been factored into the UNFCCC and Agenda 21. Chapter 2 examined the UNFCCC in terms of its reference to energy and energy access for the poor, and this section focuses on reviewing the principal non-binding outcome document that emerged from UNCED – Agenda 21 – for its specific references to the energy-poverty-climate change nexus.

Agenda 21, comprising 40 chapters, remains one of the UN's most voluminous outcome documents dealing with sustainable development. This outcome of the Rio Earth Summit is officially recorded in the UN as a five-volume document: The Rio Declaration on Environment and Development and section I ("Social and economic dimensions") of Agenda 21 are in volume I; section II ("Conservation and management of resources for development") of Agenda 21 is in volume II; and sections III ("Strengthening the role of major groups") and IV ("Means of implementation") of Agenda 21 and the non-legally binding authoritative statement of principles for a global consensus on the management, conservation, and sustainable development of all types of forests are in volume III. The proceedings of the Conference and opening and closing statements are in volume IV, while the statements made during the Summit Segment are in volume V (UNGA, 1992).

Despite the UN's past work on energy for sustainable development, and the series of 11 intergovernmental negotiating committees (INCs) related to climate change in the 1990s, the 1992 Agenda 21 – the widely hailed output of the Earth Summit – does not contain a specific chapter focusing on energy, although the issue of "energy" is referenced no less than 157 times in the entire document. Compared with the WCED report, which laid out linkages between climate change and energy, Agenda 21 included a more diffused approach to energy, with a number of references contained in Chapter 4, entitled "Changing consumption patterns", and most of the references contained in Chapter 9, entitled "Protection of the atmosphere". Curiously, the references to energy usage in developing countries and the only references to the need for developing countries to increase energy to meet their development needs are not primarily contained in either Chapters 4 or 9, but instead are found in Chapter 7 dealing with human settlements. It is therefore possible to argue that in the context of Agenda 21, energy for development issues are principally framed in terms of human settlements, whereas the energy-related adverse effects on the atmosphere are separately focused on in Chapter 9.

A review of the Agenda 21 chapters reveals that the chapters do not contain any specific references to "energy access" and "energy for sustainable development". Chapter 2, entitled "International cooperation to accelerate sustainable development in developing countries and related domestic policies", Chapter 3, more simply entitled "Combating poverty", and Chapter 5, entitled "Demographic dynamics and sustainability", contain no references whatsoever to either energy or climate change and there is no mention of energy access for the poor in any of these chapters, despite their stated focus on sustainable development in developing countries.

Energy is referenced for the first time in Chapter 4, which focused on changing consumption patterns, but again climate change is not mentioned at all in this chapter. In this chapter there are seven references to energy, four of which are contained in paragraph 18, but here the focus in not on increasing energy access but on encouraging efficiency in the use of energy and resources. The main thrust of this chapter is to frame energy in terms of how energy consumption and production patterns and attendant lifestyle choices impact the environment, but not in terms of how energy relates to poverty reduction and/or improving human well-being. An examination of the actual references to energy contained in Chapter 4 indicate that the majority of the recommendations range from considering how economic growth can occur while reducing the use of energy and material (UNGA, 1992, Chap. 4, para. 10.d); to encouraging greater efficiency in the use of energy and resources (Chap. 4, chapeau to para. 18); reducing the amount of energy and material used per unit in production of goods and services (Chap. 4, para18); and recognizing that price stimulus and market signals make the environmental cost of consumption of energy clear to producers and consumers. (Chap. 4, para. 24). Chapter 6 – "Protecting and promoting human health" – contains four references to energy but three of these are contained in subparagraph 41.j dealing with reducing the health risks associated with industry and energy production. What is interesting to note, however, is that this subparagraph contains a separate reference to indoor air pollution which calls for methods to reduce indoor air pollution and for the development of health education campaigns in developing countries "to reduce health impacts of domestic use of biomass and coal".

By way of contrast, energy is mentioned 38 times in Chapter 7, dealing with human settlement development, but again climate change is not mentioned. The bulk of the references to energy – 27 of them – are contained in a subsection of the chapter focused on sustainable energy and transport systems in human settlements from paragraphs 46–54. The stated objectives of this section are to focus on the "provision of energy-efficient technologies and alternative/renewable energy for human settlements" and "to reduce the negative impacts of energy production and use on human health and on the environment" (UNGA, 1992, Chap. 7, para. 49). Developing countries are called upon to undertake various activities that are seen as supporting these objectives, namely: reforestation programs for renewed biomass programs with "a view towards achieving sustained provision of the biomass energy needs of the low-income groups in urban areas and the rural poor, in particular women and children"; national action programs related to renewable energy sources (solar, hydro, wind and biomass); dissemination and commercialization of renewable energy technologies; and information and training programs related to energy efficiency measures and appliances (Chap. 7, para. 51). But other than the need for forest regeneration to service the energy needs of the poor, there are no explicit linkages made between increasing energy services and poverty reduction in this subsection. The only explicit references to poverty are made in relation to social and employment measures that address urban settlements.

It is in Chapter 7, dealing with human settlements, that the dichotomy faced by developing countries in terms of needing to increase energy is most clearly referenced. However, the action proposed below does not match the scope of the challenge, especially in terms of rural areas where the bulk of the poor who lack energy services reside:

> Developing countries are at present faced with the need to increase their energy production to accelerate development and raise the living standards of their populations, while at the same time reducing energy production costs and energy -related pollution. Increasing the efficiency of energy use to reduce its polluting effects and to promote the use of renewable energies must be a priority in any action taken to protect the urban environment" (Chap. 7, para. 46).

But neither this paragraph nor this segment of Agenda 21 contains any reference to the notion that developing countries will need increased access to energy resources to address poverty concerns. Additionally, both in this paragraph, and in a following paragraph focusing on transport, which lists numerous problems related to transportation in developing countries, there is no reference to climate change or fossil fuel-related emissions (para. 48). The explicit aim of this section of Agenda 21's linkage between human settlements and energy is to extend energy efficiency and renewable energy for human settlements, but it is striking that there is no explicit mention of climate change in this section.

Chapter 9 of Agenda 21 begins with the recognition that energy is essential to socio-economic development, but the dual focus of this section is primarily on the role of energy efficiency and on promoting the research, development, and use of environmentally sound energy sources, systems, and services, including, in particular, renewable energy. Again there is no direct reference to the need to increase access to energy and/or any linkages between energy and poverty reduction in developing countries, as well as no direct reference to the linkages between climate change and energy access for the poor.

The basic and ultimate objective of the energy progam proposed in this section is:

> to reduce adverse effects on the atmosphere from the energy sector by promoting policies or programmes, as appropriate, to increase the contribution of environmentally sound and cost-effective energy systems, particularly new and renewable ones, through less polluting and more efficient energy production, transmission, distribution and use" (UNGA, 1992, Chap. 9, para. 9.11).

The caveat here is that due consideration should be paid, in so doing, to the situation of countries dependent on fossil fuel generation incomes and/or use and the situations of countries that are highly vulnerable to the adverse effects of climate change. The list of specific activities that governments could undertake in cooperation with the UN and other civil society actors that would support the overall energy-related objectives of Agenda 21 are clustered primarily around energy efficiency and the development and use of new and renewable energy with the addition of a reference promoting the development of integrated "energy, environment and economic policy decisions for sustainable development" (Chap. 9, para.12), but again without any clear reference to concerns related to poverty reduction or energy access for the poor.

Finally, in terms of reviewing Agenda 21 for any stated references and linkages to the energy–poverty–climate change nexus, it is important to highlight the point that the issue of climate change is referenced predominantly in Chapters 17 and 18 dealing with marine environments and freshwater resources in terms of sea level rise and coastal and marine resources. The issue of climate change is referenced over 33 times in Agenda 21, but in fact, only three of these references are found in Chapter 9 and only one of these three references is an actual reference to the UNFCC (Chap. 9, para. 2), whilst two do not directly reference the linkage between energy access and climate change, but instead outline that "concern about climate change and climate variability, air pollution and ozone depletion has created new demands for scientific, economic and social information" to reduce uncertainties (Chap 9, para. 6) and promote energy efficiency and new and renewable energy, keeping in mind the needs of developing countries "highly vulnerable to adverse effects of climate change" (Chap. 9, para. 11). The lack of explicit linkages between energy and poverty reduction, which appears to be largely missing in Agenda 21, may be explained by the fact that the principal focus of Agenda 21 was more environmental than socio-economic. The scattershot approach to energy is in contrast to the WCED's specific chapter on energy. Climate change as an issue is treated as separate and distinct from energy issues. The lack of any sort of clearly stated linkage between energy poverty and climate change is apparent, as are the lack of references on increasing energy access for the poor.

3.3.4 Examining the energy–poverty–climate change nexus in the lead-up to, and in, the 2000 Millennium Declaration and the MDGs

In terms of energy for sustainable development, there are two important global outcomes that preceded the 2000 Millennium Declaration that need to be highlighted. The first was the UNGA's focus on energy, and the second was the publication of the *World Energy Assessment: Energy and the Challenge of Sustainability* (UNDP/UNDESA/World Energy Council [WEC], 2000).

As early as 1997, there was a broad global recognition of the importance of increasing access to energy services to address the needs of developing countries, and the "beneficial" role of increasing energy services on poverty eradication, as evidenced by UNGA's convening of a Special Session (19th) to review the 5-year follow-up to Agenda 21, which specifically stated that:

> In developing countries, sharp increases in energy services are required to improve the standard of living of their growing populations. The increase in the level of energy services would have a beneficial impact on poverty eradication by increasing employment opportunities and improving transportation, health and education. Many developing countries, in particular the least developed, face the urgent need to provide adequate modern energy services, especially to billions of people in rural areas. This requires significant financial, human and technical resources and a broad-based mix of energy sources (UNGA, 1997, para. 45).

In terms of understanding the broad intergovernmental context of energy for sustainable development, this 1997 Special Session is noteworthy because in the very next paragraph, the resolution recognizes "a need for: a movement towards sustainable patterns of production, distribution and use of energy" and states that "in order to advance this work at the intergovernmental level, the Commission on Sustainable Development will discuss energy issues at its ninth session (para. 46). Within the intergovernmental context, this would serve as the first call for global negotiations on energy for sustainable development to be held in the context of the UN's CSD.

The second influential outcome, as noted earlier, was the publication of a global energy review that provided a comprehensive analytical framework linking the global energy and sustainable development challenges, entitled *World Energy Assessment: Energy and the Challenge of Sustainability* (UNDP/ UNDESA/WEC, 2000). This report became a benchmark assessment that focused on linkages between energy and a wide range of sustainable development concerns, including poverty and gender inequality. Produced by a diverse set of globally recognized energy experts, including two of the world's leading energy experts at the time – J. Goldemberg (Brazil) and A. Reddy (India) – this energy assessment was intended to bring together energy issues relevant to both developed and developing countries and designed to be an input to the anticipated 2002 World Summit on Sustainable Development (WSSD). The scope and process of putting together such an assessment on energy remain unprecedented to this day, in that the assessment began in 1998 and included an outreach effort by each of the three founding organizations, 12 convening lead authors, each of them with associated teams of experts, and a lengthy and global editorial review process, including inputs from governments and civil society actors. The end result was a 489-page report that would go on to become a reference and policy tool in the lead-up to intergovernmental negotiations for the 2002 WSSD and would become the template for more recent efforts to put together the Global Energy Assessment spearheaded by the International Institute for Applied Systems Analysis (IIASA).

Chapter 2 ("Energy and social issues") of the *World Energy Assessment* provided a clear and concise linkage stating that poverty alleviation and development depend on universal access to energy services that are affordable, reliable, and of good quality, which is a sentiment that has been referenced by various developing countries in the CSD and the 2002 WSSD negotiations, and continues to resonate in terms of negotiation stances of developing countries as well as in a variety of other global initiatives, including the SE4All. Chapter 2 also provided the first globally relevant and best-known definition of the term "energy poverty":

> The energy dimension of poverty—energy poverty—may be defined as the absence of sufficient choice in accessing adequate, affordable, reliable, high-quality,

safe, and environmentally benign energy services to support economic and human development. The numbers are staggering: 2 billion people are without clean, safe cooking fuels and must depend on traditional biomass sources; 1.7 billion are without electricity (UNDP/UNDESA/WEC, 2000, p. 44).

The UN's historic Millennium Declaration emerged as a result of a special 3-day summit convened at the UN headquarters in NY. In all, 147 heads of state and government, constituting the largest grouping of world leaders ever present, adopted the Millennium Declaration on September 8, 2000. A total of 189 countries signed up to the Millennium Declaration – which was, and remains, the only globally agreed commitment to halve extreme poverty and achieve equitable and sustainable development for all. The Declaration was categorical and sweeping in its scope and the tone highlighted the relationship between development and poverty eradication, stating:

> We will spare no effort to free our fellow men, women and children from the abject and dehumanizing conditions of extreme poverty, to which more than a billion of them are currently subjected. We are committed to making the right to development a reality for everyone and to freeing the entire human race from want (UN, 2000, para. 11).

The nexus between sustainable development and poverty reduction was prominently highlighted, in that the Declaration expressed concern about the obstacles faced by developing countries in financing their sustainable development. It called for the cancellation of "all official bilateral debts of those countries in return for their making demonstrable commitments to poverty reduction", and tied the grants of "more generous development assistance especially to countries that are genuinely making an effort to apply their resources to poverty reduction (UN, 2000, para. 11). The Declaration announced the creation of "an environment – at the national and global levels alike – which is conducive to development and to the elimination of poverty" (para. 12). It also highlighted the need to address gender equality as a way to combat poverty and the role of civil society partnerships to promote action on development and poverty eradication (para. 20). The remaining references linking poverty eradication and sustainable development are contained in the section of the Declaration dealing with the special needs of Africa in paragraphs 27 and 28, and the final reference is contained in the segment dealing with the global efforts to make the UN "a more effective instrument for pursuing" a variety of priorities, such as the fight for development, fight against poverty, ignorance and disease, the fight against injustice, violence etc. (UN, 2000, para. 29).

And yet, energy is not mentioned or referenced even once in the entire nine-page Declaration. The issue of energy being the driver of human development and of the role of energy being crucial for meeting basic human needs – food, water, shelter – and addressing poverty is entirely absent from the Declaration. Notwithstanding the 1997 Special Session of the UNGA's own recognition of the linkage between increasing energy access and poverty reduction, and the widely disseminated work of the influential *World Energy Assessment* released prior to the Summit, the linkages between energy and poverty reduction were not explicitly mentioned in either the 2000 Millennium Declaration or the 2001 MDGs. Although there is no specific mention of the nexus between energy and climate change, the Declaration called for "every effort to ensure the entry into force of the Kyoto Protocol, preferably by the tenth anniversary of the United Nations Conference on Environment and Development in 2002, and to embark on the required reduction in emissions of greenhouse gases" (UN, 2000, para. 23). The heart of the Declaration that would be translated into key MDGs is contained in the section on "Development and Poverty Eradication" highlighted in Box 3.4.

As a follow-up to the Millennium Summit, an influential report was prepared by the UN Secretary-General to guide the actual and overall implementation of the Millennium Declaration (UNGA, 2001). This 55-page report, entitled *Road Map Towards the Implementation of the United Nations Millennium Declaration*, contained no fewer than 56 goals, ranging from peace and security issues to combating HIV in the main body of the report. As stated in the introduction of the report, the preparation of the report and the collation of a set of eight MDGs in a separate Annex were a result of consultations held amongst the UN family of agencies and other intergovernmental organizations (International Monetary Fund [IMF], OECD, and the World Bank). The stated intent behind having a limited set of eight goals was that "clear and stable numerical targets can help trigger action and promote new alliances for development" (UNGA, 2001, p. 55). The Annex lists a set of eight goals, 18 targets and 48 indicators. None of the eight goals or 18 targets reference energy in any way, and only one indicator (indicator 27) associated with goal 7 (ensure environmental sustainability) is listed as relating to energy, namely "GDP per unit of energy use (as a proxy for energy efficiency) (UNGA, 2001, p. 57).

A review of the road map report reveals that there is only one reference to energy, which is listed as one of a series of issues, including the management of water and biodiversity resources in connection with concerns of small island developing states (SIDS). Part IV of the report, entitled "Protecting our common environment", has several paragraphs focusing on the need to ensure the entry into force of the Kyoto Protocol, but here the only reference to energy is found in paragraph 170 in connection with the anticipated 2002 WSSD having a focused agenda that would "foster discussion of … in particular environmental sectors (forests, oceans, climate, energy, fresh water, etc.)" (UNGA, 2001, p. 32).

Box 3.4 Millennium Declaration: Key resolved actions that would be translated into Millennium Development Goals (MDGs)

Paragraph 19

We resolve further:

To halve, by the year 2015, the proportion of the world's people whose income is less than one dollar a day and the proportion of people who suffer from hunger and, by the same date, to halve the proportion of people who are unable to reach or to afford safe drinking water.

a. To ensure that, by the same date, children everywhere, boys and girls alike, will be able to complete a full course of primary schooling and that girls and boys will have equal access to all levels of education.
b. By the same date, to have reduced maternal mortality by three quarters, and under-five child mortality by two thirds, of their current rates.
c. To have, by then, halted, and begun to reverse, the spread of HIV/AIDS, the scourge of malaria and other major diseases that afflict humanity.
d. To provide special assistance to children orphaned by HIV/AIDS.
e. By 2020, to have achieved a significant improvement in the lives of at least 100 million slum dwellers as proposed in the "Cities Without Slums" initiative.

Paragraph 20

We also resolve:

f. To promote gender equality and the empowerment of women as effective ways to combat poverty, hunger and disease and to stimulate development that is truly sustainable.
g. To develop and implement strategies that give young people everywhere a real chance to find decent and productive work.
h. To encourage the pharmaceutical industry to make essential drugs more widely available and affordable by all who need them in developing countries.
i. To develop strong partnerships with the private sector and with civil society organizations in pursuit of development and poverty eradication.

Source: UNGA (2000).

From a broader sustainable development perspective, the most important part of the report was actually contained in the sole Annex, entitled "Millennium Development Goals", which listed a series of eight goals drawn exclusively from the section of the Millennium Declaration dealing with development and poverty eradication. To be clear, the 56 goals contained in the body of the actual report included these eight goals, but it was only this particular set of eight goals that were distinguished from the rest and constituted to form the MDGs (Figure 3.2). While access to energy services as a crucial foundation for development was broadly recognized by the UN as an important policy intervention for meeting the MDGs, these goals lack any specific energy-related target or policy-specific guidance (UNDP, 2005).

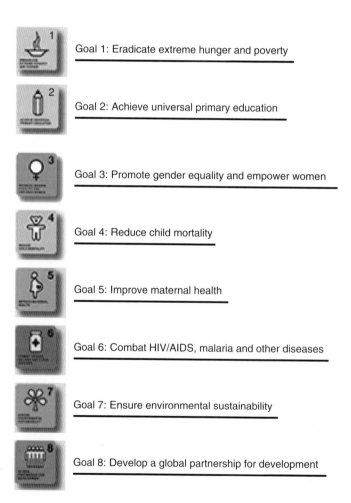

Goal 1: Eradicate extreme hunger and poverty

Goal 2: Achieve universal primary education

Goal 3: Promote gender equality and empower women

Goal 4: Reduce child mortality

Goal 5: Improve maternal health

Goal 6: Combat HIV/AIDS, malaria and other diseases

Goal 7: Ensure environmental sustainability

Goal 8: Develop a global partnership for development

Figure 3.2 The eight Millennium Development Goals (source: Millennium Project website: www.unmillenniumproject.org/goals).

3.4 Examining the energy–poverty–climate change nexus in key post-millennium UN sustainable development outcomes (2001–2012)

The trend of carrying out a 5-year review after the 1992 Rio Summit, followed by the convening of a series of preparatory committees, leading on to a massive global conference after 10 years resulted in the 2002 WSSD held in Johannesburg, South Africa. Ten years after the 2002 WSSD, the UN again convened a series of preparatory meetings that would lead to the 2012 UN Conference on Sustainable Development – otherwise also known as the Rio+20 Summit. This section of the chapter examines the post-2000 outputs of the CSD process, the 2002 Johannesburg Plan of Implementation (JPOI) and the Rio+20 outcome document entitled *The Future We Want* in terms of any stated references and concrete linkages on energy access for the poor and climate change.

Energy for sustainable development issues was considered as a stand-alone thematic issue for the first time in the UN context under the aegis of the UN's CSD. In order to advance work on energy, the 1997 UNGA Special Session adopted a multi-year program of work for the CSD for the period 1998–2001, under which the sectoral theme of the ninth session (2001) of the CSD was atmosphere/energy and the economic sector to be discussed was energy/transport. Recognizing the complexities and interdependencies inherent in addressing energy issues within the context of sustainable development, the General Assembly decided that preparations for the ninth session of the CSD (CSD-9) should be initiated at the seventh session of the CSD (CSD-7) and should utilize an Open-Ended Intergovernmental Group of Experts on Energy and Sustainable Development to be held in conjunction with intersessional meetings of CSD-8 (2000) and CSD-9. In order to facilitate coordination and cooperation among UN entities in energy-related areas, an Ad Hoc Inter Agency Task Force on Energy was established in February 1998. This task force decided on specific inputs from the UN system that would contribute to the work of the Open-Ended Intergovernmental Group of Experts on Energy and Sustainable Development and the CSD. In addition, the Committee on Energy and Natural Resources for Development was established by the Economic and Social Council Resolution 1998/46 of July 31, 1998 comprising government-nominated experts divided into two sub-groups of 12 members each, dealing with issues related to energy and water resources, respectively.

3.4.1 Examining the energy–poverty–climate change nexus in the 2001 CSD

The 2001 CSD-9 constituted the first intergovernmental, UN-led forum in which both intergovernmental and multi-stakeholder attention was paid to energy as a distinct and stand-alone sustainable development issue. But, more significantly, it was also the first time that energy was discussed in terms of its specific linkages to broader global sustainable development concerns and objectives within the UN framework. In addition to the inputs from the two Ad Hoc Task Forces and the *World Energy Assessment* 2000, the UNDESA convened a series of regional consultative meetings and commissioned global experts to prepare regional background papers that would provide an opportunity for national governments and relevant regional energy experts to give feedback on energy-related issues that are deemed critical for each region, so there was a regional emphasis on energy that recognized diversity of energy needs and concerns. At CSD-9, delegates focused their attention on a series of seven key energy-related issues:

- Accessibility of energy
- Energy efficiency
- Renewable energy
- Advanced fossil-fuel technologies
- Nuclear energy technologies
- Rural energy
- Energy and transport.

A number of overarching or cross-cutting issues were also considered at CSD-9:

- Research and development
- Capacity-building
- Technology transfer
- Information sharing and dissemination
- Mobilization of financial resources
- Making markets work more effectively for sustainable development
- Multi-stakeholder approach and public participation.

The manner in which energy and climate change issues are dealt with in the outcome report of CSD-9 should be contrasted with previous UN global conferences and agreed outputs – UNCHE, WCED, and Agenda 21. In terms of the final report of CSD-9, issues such as energy for sustainable development and protection of the atmosphere, which included references to climate change, transport, information for decision-making, and international cooperation, were each considered separately. Accordingly, there were separate and different decisions arrived at in terms of each of the issues: energy (Decision 9/1), protection of the atmosphere (Decision 9/2), transport (Decision 9/3) and so on (UNECOSOC, 2001).

There are 17 references to "climate change" in the CSD-9 outcome report, but the issue of "climate change" was actually only referenced three times in conjunction with the decision on the protection of the atmosphere (Decision 9/2), but not at all in the decision on energy (Decision 9/1). This is an important point to highlight because it is in keeping with the UN trajectory of addressing climate change as separate and distinct from addressing energy for sustainable development. So, for instance, the first time climate change is mentioned is in the third paragraph of the decision focused on protection of the atmosphere. But here, the decision does not emphasize or highlight anthropogenic climate change as a key global concern or challenge, but instead strikes a more measured tone, noting that:

> human activities and natural disasters contribute to the build-up of atmospheric substances, which has implications for climate change and climate variability, for the depletion of the stratospheric ozone layer and for air pollution, in particular trans boundary, urban and indoor air pollution (UN, 2001, p. 16).

In other words, anthropogenic activities and natural disasters are clustered together for the first time and seen as jointly contributing to the growth of atmospheric substances which are not directly linked to climate change, but instead have "implications" for climate change, ozone depletion, and air pollution. In this CSD-9 report, the urgency of anthropogenic-driven climate change was not mentioned and does not include any energy-related linkages. Additional references to climate change included in this decision relating to atmospheric protection are in connection to the need for cooperation across global conventions, including the UNFCCC (UN, 2001, p. 17) and a call for the Intergovernmental Panel on Climate Change (IPCC) to support the participation of developing country experts in its work (UN, 2001, p. 18).

The ninth session of the CSD (CSD-9) also included a continuation of the CSD's unique tradition of having both a multi-stakeholder dialogue and a high-level segment, which meant that both civil society actors and government ministers had a forum, albeit different in scope and reach, in which to focus on energy issues related to sustainable development. Based on a previously established format, the overall report of the CSD-9 included a Chairman's summary on key outcomes of both the multi-stakeholder dialogue (with four set themes) and the high-level segment.

The bulk of the remaining references to climate change are contained in the segment of the CSD-9 report dealing with the Chairman's summary of key outcomes of the stakeholder dialogue. The first mention of climate change is in relation to the NGO Caucus for Climate Change and Energy which is one of the stakeholder groups that received multiple references related to climate change in CSD-9 report. So even though there was such a caucus linking both issues, the actual contents of the report and its separation of climate change and energy for sustainable development decisions demonstrate the intergovernmental preference for handling energy and climate change as separate issues.

This section contained several references to the need for "the ratification of the Kyoto Protocol before the Johannesburg 2002 " given "some expressions of disappointment" with the UNFCCC-led negotiations at the sixth COP (UN, 2001, para. 22 and 25, pp. 47–48). The remaining references to climate change are contained in the section dealing with sustainable transport, including increasing demand of private vehicles causing, amongst other things, unsafe traffic, political conflict, and climate change (UN, 2001, para 40, p. 41). Meanwhile, in the section dealing with the Chairman's summary of the high-level segment – which is where the official intergovernmental aspects of the CSD are emphasized – the issue of climate change is mentioned directly only once in the last paragraph, dealing with the Chair's summary of the intergovernmental inputs on atmosphere, but the summary also notes that SIDS representatives felt that the Kyoto Protocol was inadequate in terms of level of ambition and emissions targets but still was recognized as a necessary first step (UN, 2001, p. 49).

The ninth session of the CSD was one of the most significant intergovernmental negotiations in terms of energy for sustainable development, in particular the issue of accessibility of energy or energy access, even though it stayed studiously away from any explicit linkages between energy and climate change. Decision 9/1 of the CSD-9, entitled "Energy for sustainable development", began with the obvious "Energy is central to achieving the goals of sustainable development" and then went on to focus on the "magnitude and scale of energy needs" in relation to sustainable development, which was "gauged by the fact that nearly one third of the global population of six billion, mostly living in developing countries, continue to lack access to energy and transportation services" (UN, 2001, p. 1). The Decision 9/1 was also distinctive in that the issue of accessibility of energy was accorded importance and referenced as the first of five key issues that were seen as crucial to the broader global discussion on energy for sustainable development. But the issue of energy access is referenced just once in this decision in relation to the overall recommendations related to energy access, which included the establishment and strengthening of "national and regional arrangements for promoting energy accessibility within the country" (UN, 2001, p. 4).

There are four references to energy access contained in the paragraphs 5, 6, 7, and 13 of the Chairman's summary of the multi-stakeholder dialogue on sustainable energy and transport. In this section, the summary of opinions and views expressed by state and non-state actors included the following references to energy access: scientists pointed out that "developed country energy access was defined by the spread of markets" (para. 5); workers and trade unions "highlighted the interaction with and social impact of energy access policies on employment" (para. 6); local authorities called for more investment in developing decentralized energy systems and "establishing various efficiency and design initiatives to improve energy access at the local level" (para. 7);

and trade unions called for a "continued public sector role in energy access" whilst scientists "said that a services approach could address access without compromising lifestyle choices" (para. 13) (UN, 2001, pp. 34–35).

The Chairman's summary of the CSD-9's high-level segment (governments had official, ambassadorial, or ministerial representation) is a key outcome in the overall examination of the intergovernmental process on energy. Paragraph 8 the Chair's summary CSD-9 report contained in the section dealing with energy for sustainable development stands out. It references the fact that the main challenge "continues to be the provision of energy services to 2 billion people in developing countries currently without access to such services", but it is what comes after this statement that it is important to highlight. It is the first ever mention of a possible global target related to energy access: "In this context, some proposed that the World Summit on Sustainable Development should adopt a target of cutting by half by 2015 the proportion of people without access to clean fuels and electricity" (UN, 2001, p. 46).

This is the first mention in any UN-led global negotiations of an energy access target, even though in this case it is contained in the Chair's summary and not in any actual negotiated decision. The following paragraph, paragraph 9, is explicit in pointing out that the "main goal of energy for sustainable development should be poverty eradication" (UN, 2001, p. 46). It is worth noting that there were no clear statements or references to an energy access target in previous global outcomes, and as the discussion on the WSSD's outcome – the JPOI – demonstrates, this target would not be mentioned in 2002 JPOI .

Unlike the actual negotiated decision on energy for sustainable development, the very next paragraph (para. 9) of the report leaves no doubt that the "main goal of energy for sustainable development should be poverty eradication" and that "international efforts to achieve this goal should be guided by the principle of common but differentiated responsibilities" (UN, 2001, p. 46). The role of access to energy being "crucial to economic and social development, and alleviation of poverty" and the idea that "lack of access to energy is the main cause of poverty" are again mentioned in paragraph 10 (UN, 2001, p. 46). These clear and reiterated references linking energy access in developing countries to poverty eradication stands in contrast to the lack of any linkages to energy services, poverty reduction and sustainable development in both the Millennium Declaration and the MDGs.

3.4.2 Examining the energy–poverty–climate change nexus in the 2002 JPOI

CSD-9 recommendations on energy played a role in informing the intergovernmental negotiations, including the preparatory process leading up to the 2002 WSSD, and its principal output, the JPOI. The 2002 WSSD was a massive UN summit, which was attended by 21,000 people, including 9,101 delegates, 8,227 NGO representatives, and 4,012 accredited media (UN News Centre, 2002a). As in the case of previous UN global conferences, preparatory negotiations leading up to the 2002 WSSD took 9 months and spanned three continents. In the lead-up to the WSSD, the UN Secretary-General, Kofi Annan, initiated a detailed process of evaluation led by the UNDESA but comprising inter-agency inputs from all relevant UN agencies on five thematic areas – water, energy, health, agriculture, and biodiversity (WEHAB). As part of the WEHAB initiative in the lead-up to the Johannesburg Summit, five thematic papers were prepared following the consultative process within the UN system. The papers were not intended to be consensus documents reflecting the totality of UN system activities in the five areas; rather their aim was to provide a broad overview of each thematic area, to highlight linkages among the areas, identify key gaps and challenges, and to highlight issues for further action. This idea of linkages between each of the thematic topics and the list of action items for future action in the WEHAB papers comprised the first globally relevant attempt to look at the nexus amongst key sustainable development concerns.

The energy paper produced as part of the WEHAB initiative, entitled *A Framework for Action on Energy*, remains one of the first and only UN-commissioned reports on energy for sustainable development that examined energy specifically in relation to four other cross-cutting development themes – water, health, agriculture, and biodiversity. It also provides a framework for linking energy-related concerns with poverty reduction and climate change challenges by noting up front that the lack of access to diverse and affordable energy services "correlates closely with many indicators of poverty, such as poor education, inadequate health care and hardships imposed on women and children", whilst at the local and national levels, reliable energy services are key to economic stability, growth, jobs, and living standards (UN, 2002, p. 7). The report is clear in elaborating the linkages between energy, climate change, and ill-health related to air pollution, pointing out that:

> …the combustion of fossil fuels is the largest source of health-damaging air pollutants, as well as being the major source of greenhouse gas (GHGs) emissions. The emission of fine particulate matter – from the burning of coal, oil, diesel fuel, gasoline and wood in transportation, power generation and space heating – can lead to respiratory problems and cancer. Indoor fires burning coal, wood or other biomass fuels are also a significant source of particulate pollution in rural homes. Smoke from cook fires contains dangerous amounts of toxic substances and can also lead to respiratory problems (UN, 2002, p. 8).

The WEHAB paper on energy also clearly referenced the fact that the growing demand for energy services in developing countries constituted an opportunity to meet energy demands in sustainable ways, and noted that "negative effects of climate change will be most severe in developing countries and be felt most by poor people" (UN, 2002, p. 9). In addition to focusing on the linkages between energy and the other four thematic issues, it went further and also discussed energy in terms of the five cross-cutting issues that had previously been identified by CSD, namely: accessibility; energy efficiency; renewable energy; advanced fossil fuel technologies; and energy and transport. Finally, unlike other UN contributions, the paper also includes a detailed framework for action, with by far the most comprehensive list of 12 action areas for energy, for which targets, examples of activities, and actions to address cross-cutting issues were listed.

As Box 3.5 indicates, no other UN-sponsored paper or outcome, including those emanating from the WEO and more recently convened initiatives such as the SE4All initiative, include or provide the level of detail on indicative targets that was provided in this 2002 background conference paper. Twelve action areas and their associated indicative targets were provided in 2002 but somehow failed to gain traction in the intergovernmental process. More than a decade later, in 2014, the idea that more basic energy for SDGs and targets associated with the SDGs can seem somehow novel or new to the intergovernmental process is hard to explain and rationalize.

Despite the submission of WEHAB initiative papers prior to the convening of the WSSD, the JPOI did not contain any references to energy-related targets. Unlike CSD-9, whose recommendations were not binding, the JPOI's energy-related recommendations constitute the first globally agreed intergovernmental references related to energy for sustainable development, including energy and transport. The JPOI, like Agenda 21 before it, clearly states that eradicating poverty was the "greatest global challenge facing the world" and recognized that poverty eradication is "an indispensable requirement for sustainable development, particularly for developing countries" (UN, 2002, para. 7, p. 13).

As in the case of Agenda 21, the JPOI does not contain a stand-alone section or chapter for energy, with references to energy scattered through it and contained in several sections. They are primarily clustered in two paragraphs, namely paragraph 9 of section II ("Poverty eradication") and paragraph 20 of section III ("Changing unsustainable patterns of consumption and production"). Paragraph 9 focuses primarily on access to energy services in addressing poverty reduction. A call on governments to implement the conclusions adopted by the CSD-9 (para. 20) contains an exhaustive list of 23 proposed actions, but none references any concrete linkages with increasing energy access for the poor, and instead reference some of the following issues:

- increased role of energy efficiency;
- development and dissemination of alternative energy technologies;
- increased use of renewable energy;
- development and use of indigenous energy sources;
- research and development and transfer of energy technologies, including advanced and cleaner fossil fuel technologies;

Box 3.5   List of action areas and associated indicative targets: WEHAB (water, energy, health, agriculture, and biodiversity) initiative's Framework for Action in Energy

Action Area 1: Reduce poverty by providing access to modern energy services in rural and peri-urban areas.
Indicative target: To achieve the MDG of reducing the proportion of people living in extreme poverty by half, commensurate decreases in the number of people without access to electricity and clean cooking fuels are required. This implies targeting 800 million to 1 billion people to be provided with modern energy services by 2015. This corresponds to half of the estimated number of people currently living in extreme poverty.
    Appropriate intermediate targets to achieve this are required.

Action Area 2: Improve health and reduce environmental impacts of traditional fuels and cooking devices.
Indicative target: The 400 million households that currently depend on traditional fuels need access to modern efficient cooking fuels and systems. This will contribute to addressing gender inequity at the household and community level. Appropriate intermediate targets are required to achieve this.

Action Area 3: Improve access to affordable and diversified energy sources in Africa.
Indicative target: Substantially increase access to modern energy services from an estimated baseline situation of 10 per cent of the population in rural sub-Saharan Africa. Develop appropriate institutional and regulatory frameworks.

Action Area 4: Reduce poverty by providing access to modern energy services in rural and peri-urban areas.
Indicative target: In order to realize the potential of end-use energy efficiency improvements, which are estimated to be in the range of 25–40 per cent in residential and commercial buildings, industry, agriculture and transport sectors in all countries, appropriate targets for every five years are needed.

Action Area 5: Improve energy efficiency in all sectors using established practices on standards and labeling techniques.
Indicative target: Substantially increase the application of appropriate energy efficiency standards and labelling programmes from the current coverage in about 30 developing countries to a much larger number.

Action Area 6: Improve efficiency in power generation.
Indicative target: To improve the efficiency of converting fuels to power from the current low levels, it is necessary to substantially increase the share of modern electricity generation technologies, such as natural gas-based combined cycle, in national supply mixes.

Action Area 7: Progressively increase contribution of renewable energy mix of all countries.

Indicative target: Progressively increase the contribution of renewable energy in the global primary energy mix from the current base line of 2 per cent for modern renewables. For example, at the current rate of expansion, wind energy is expected to increase from the existing generating capacity of 25,000 MW to 100,000–150,000 MW in the next decade. Targets are required to generate similar trends in other forms of renewable energy such as biomass, solar, hydro and biofuels.

Action Area 8: Improve access to basic health care and education for poor people through the provision of renewable energy systems in primary health care centres and schools.

Indicative target: At 1–2 kW per health care centre, 100–200 MW capacity is required for 100,000 health care centres (vaccine refrigerators, water pumps and other allied health systems). At 500 W per school, 100,000 schools require 50 Mw capacity. (Particular focus on rural and remote areas).

Action Area 9: Promote the use of renewable energy in vaccine and immunization programmes.

Indicative target: To support the achievement of the MDG on reducing under-5 mortality by two-thirds, provide all vaccine and immunization programmes and centres with appropriate renewable energy systems (to suit local conditions). The average vaccine refrigerator requires 250–500 W.

Action Area 10: Provide the use of renewable energy to facilitate access to safe drinking water.

Indicative target: To support the achievement of the MDG to reduce by half the proportion of people who do not have access to safe drinking water, it would be necessary to reach 500 million people with 40 litres per capita, which would require 1 million water pumps. (One pump is expected to serve on average 500 people in a community).

Action Area 11: Increase the use of advanced fossil fuel technologies for energy generation.

Indicative target: Assuming that capital stock renovation of 5–10 per cent a year can be achieved, the entire energy system can be upgraded with advanced technology options in the next 20–30 years if performance criteria are explicit.

Action Area 12: Promote the use of clean coal technologies (CCTs) in countries using coal.

Indicative target: Given current technology availability and trends, starting from 2005, 12 GW per year of clean coal technologies in the next 10 years is feasible.

Source: UN/WEHAB (2002, p. 16).

- networking between centres of excellence on energy for sustainable development;
- improving the functioning, transparency, and information about energy markets;
- phasing out of relevant subsidies that inhibit sustainable development;
- cooperation between international and regional institutions dealing with energy;
- regional cooperation arrangements for promoting cross-border energy trade;
- facilitation of dialogue forms among regional, national, and international producers of energy.

The main paragraph dealing with improving access to reliable and affordable energy services for sustainable development in the JPOI – paragraph 9 – does not contain a direct energy-related target, but instead there is more generalized recognition of the need for access to reliable and affordable energy services "sufficient" to facilitate the achievement of the MDGs. Drawing directly from CSD-9, this paragraph of the JPOI specifically recognized that "access to energy facilitates the eradication of poverty" and lists seven sub-paragraphs of actions at all levels. But these series of "actions" did not contain any targets, unlike the targets ascribed to the provision of clean drinking water and adequate sanitation in the paragraph immediately preceding the one on energy. Moreover, what is striking is that any reference on the limited access to and use of traditional biomass and inefficient energy services impacting negatively on the poor, including health and gender, are missing in this paragraph dealing with energy. Instead of a specific target focusing on halving, for example, the proportion of people who lack access to basic clean energy services, as was done in the case of clean water, the JPOI focuses on instead on seven broad and generalized aspirational needs for instance:

a. Improve access to reliable, affordable, economically viable, socially acceptable, and environmentally sound energy services and resources taking into account national specificities and circumstances through various means … including through capacity building, financial and technological assistance and innovative financing mechanisms, including at the micro-and meso-levels, recognizing the specific factors for providing access to the poor…

b. Promote a sustainable use of biomass and, as appropriate, other renewable energies through improvement of current patterns of use, such as management of resources, more efficient use of fuelwood and new or improved products and technologies…

c. Assist and facilitate on an accelerated basis, with financial and technical assistance of developed countries, including through public and private partnerships, the access of the poor to reliable, affordable, economically viable, socially acceptable and environmentally sound energy services … that energy services have positive impacts on poverty eradication and improve the standard of living" (UN, 2002, pp. 11–12).

It turns out that the programmatic recommendation on energy access for the poor that is not associated with any lengthy qualifiers is not contained in the paragraph on energy access, but instead is located in the paragraph that follows on strengthening the contribution of industrial development to poverty eradication and sustainable natural resource management. Paragraph 10 contains a series of six actions and includes a need to: "provide support to developing countries for the development of safe, low-cost technologies that provide or conserve fuel for cooking and water heating" (UN, 2002, p. 17).

Although the JPOI does not provide any guidance as to the nature, scope, and framework for the provision of support by industrial development, it is this kind of action that is directly relevant to improving energy access, and thereby improving the lives of the poor, because it offers both health and environmental benefits. The JPOI contains no recommendations related to the creation or establishment of specific means of implementation, that is, any specific financing or capacity-related mechanisms for achieving various energy for sustainable development objectives listed. So while it does include lengthy paragraphs extolling the benefits of various energy for sustainable development recommendations, it is does not actually provide any concrete guidance on how these recommendations will be implemented at the programmatic level, nor does it reference any tangible means of financing for implementation.

The JPOI, like Agenda 21, does not include explicit language emphasizing global climate change and its energy linkages. In fact, the first reference to global climate change in the JPOI only occurs in a sub-paragraph in section IV ("Protecting and managing the natural resource base of economic and social development"), which mentions the "critical uncertainties for the management of the marine environment and climate change" (UN, 2002, para. 30(b) p. 29). The issue of climate change is primarily discussed in paragraph 38, which states that climate change and its adverse effects are "a common concern of humankind". Deep concern is expressed that all countries but particularly developing countries, including LDCs and SIDS, "face increased risks of negative" climatic impacts, and it is in this context that "the problem of poverty, land degradation, access to water and food and human health are seen to be central for global attention", but there is no mention of the problem of lack of energy access for the poor. The JPOI recognizes the UNFCCC as the "key" instrument for addressing climate change and calls for the timely ratification of the Kyoto Protocol but this section contains only one explicit linkage or mention of energy. The sole reference is in connection with the call to "develop and disseminate innovative technologies in regard to key sectors of development, particularly energy, and of investment in this regard" (UN, 2002, para. 38(f), p. 36). The other few remaining references to climate change arise in relation to its impacts and role in developing country groupings, such as African LDCs and SIDS.

In terms of the broader nexus between energy access for the poor and climate change, one of the unique aspects of the JPOI which continues to be emphasized is its recognition and elevation of the role of "partnerships for sustainable development". At a press conference on the final day of the WSSD Summit, UN Secretary-General Kofi Annan specifically highlighted the role of partnerships sponsored by the conference which he said marked "a major leap forward" in teaming up the public sector, civil society, businesses, and other key actors in the global fight against poverty (UN News Centre, 2002b). The JPOI's emphasis on the multi-stakeholder and partnership approach has long-term implications for civil society involvement in implementing sustainable energy and energy access issues. The role of partnerships and voluntary initiatives related to the implementation of sustainable development objectives can be traced to back to the JPOI and remains a crucial factor in the UN's ongoing quest for a post-2015 development agenda. The promotion of "partnerships" was a unique outcome of the WSSD and the effective implementation of these partnerships brings new sets of opportunities and challenges that will be discussed in Chapter 4.

Often referred to as type II voluntary initiatives (to distinguish them from the type I, or politically negotiated, agreements emanating from the WSSD), these voluntary partnerships are conceived as facilitating new and complementary forms of cooperation among diverse stakeholders. These voluntary type II partnerships have the potential to become a significant, complementary, multi-stakeholder delivery mechanism for the implementation of sustainable development objectives in general, and energy objectives in particular. But many questions remained unanswered, such as what are the specific linkages (financing modalities and substantive, institutional, and monitoring and evaluation arrangements) between type I outcomes (those contained in the JPOI) and type II partnerships; and what procedures best facilitate the broader implementation and scaling up of effective "partnerships" for sustainable development? This is particularly relevant in the case of energy and transport issues where multiple areas for action have been identified, such as promotion of renewable energy and energy efficiency, and advancing the use of cleaner fossil fuels. The JPOI's recognition of the role of "partnerships for sustainable development" is echoed in the more recent trend towards the voluntary, non-UN-driven multilateral partnership initiatives related to energy and climate change, which are discussed in the Chapter 4.

3.4.3   Examining the energy–poverty–climate change nexus in the lead-up to, and at, the 2012 Rio+20 Summit: AGECC/SE4All, Global Sustainability Panel Report and the Rio+20 Summit's Future We Want

Following the 2002 JPOI, the CSD once again became the official intergovernmental forum for follow-up on the status of implementation of the JPOI. The first crucial post-WSSD intergovernmental meeting was the CSD-11, which decided that its multi-year program of work (2004/05–2016/17) would be organized on the basis of seven 2-year cycles, with each cycle focusing on selected thematic clusters of issues. Energy for sustainable development, industrial development and air pollution along with "atmosphere climate change" were listed as the focus of the second work cycle and discussed in 2006/07 (UN, 2003). CSD-11 also agreed that each cycle would address the thematic clusters of issues in an integrated manner, taking into account economic, social, and environmental dimensions of sustainable development. The Commission further agreed that means of implementation should be addressed in every cycle and for every relevant issue, action, and commitment. Linkages to other cross-cutting issues are also to be addressed in every cycle.

The CSD's 14th session in 2006 and 15th session in 2007 focused on a cluster of thematic issues, which included energy for sustainable development, industrial development, air pollution/ atmosphere, and climate change. But right from the start, political difficulties in linking the energy and climate change policy debates at the global level were evident during the proceedings of the CSD-15. As the main policy session of the second implementation cycle, CSD-15 (2007) continued CSD-14's focus on the areas of energy for sustainable development, industrial development, air pollution/ atmosphere, and climate change.

Intergovernmental negotiations aimed at reconciling energy for sustainable development with climate change objectives were, however, fraught with tensions, and for the first time in the history of the Commission's deliberations, there was no consensus-based decision text that emanated at the May 11, 2007 conclusion of CSD-15. The result of the tense divisions amongst delegates on key points related to climate change and energy meant that the Chairman's summary text was not accepted as a consensus agreement at the concluding session of the high-level segment of CSD-15.

The Chair's summary actually states that "delegates achieved near unanimity on the industrial development and air pollution/atmosphere themes, but remained divided on key points related to climate change and energy chapters". As a consequence, it was necessary for the Chairperson to present a decision text for consideration by the Commission, reflecting the Chairperson's "best efforts to reconcile the remaining conflicting viewpoints in a fair and balanced manner" (UN, 2007, pp. 2–3). The principal reason for the breakdown in negotiations on energy for sustainable development at CSD-15 was that no consensus could be reached over the issue of energy-related targets. Recognizing the polarizing nature of "targets", the CSD-15 Chair's summary of the negotiation proceedings specifically notes:

All the major political groupings, save one, accepted the Chairman's proposed decision text. Germany, on behalf of the EU members, as well as one country attending as an observer, rejected the decision text because agreement could not be reached on time-bound targets for renewable energy, the integration of energy policies into national planning by 2010, a formal review arrangement for energy issues in the UN and an international agreement on energy efficiency (UN, 2007, p. 3).

In the end, as a consequence of not being able to reach an agreement, the final outcome of CSD-15 was the presentation of the Chairman's "summary of the negotiations" in lieu of an actual consensus-based decision document. This was a precedent, setting the CSD conclusion in terms of not being able to secure a consensus-based decision. That the lack of a global consensus came down to disagreements over the issue of how to handle energy-related targets should be underscored and highlighted for its implications for global negotiations on energy for sustainable development. Whether this "non-outcome of CSD-15" would be a harbinger of non-action in terms of future energy-related targets would remain a relevant question in light of the reluctance to formally adopt the SE4All goals within the Rio+20 outcome.

A considerable amount of attention had been paid to sustainable development indicators and targets in the lead-up to the CSD, but CSD-15 proved that global consensus on energy targets and indicators would be hard to secure. Following the decisions taken by the CSD, and Agenda 21, UNDESA began working as early as 1995 to produce an overall set of indicators for sustainable development. This effort concluded with a package of 58 indicators for sustainable development, but only three were energy-related – annual energy consumption per capita, intensity of energy use, and share of consumption of renewable energy resources (UNDESA, 2001). In order to support the efforts of the CSD, and to provide more clarity on energy-related indicators of sustainable development, the IAEA started a long-term program addressing indicators for sustainable energy development (ISEDs) in 1999. This was done in cooperation with various other international organizations, including the IEA, UNDESA, and some Member States of the IAEA.

Given the final outcome of CSD-15 negotiations, the question is whether there is the requisite global commitment to accept "soft or voluntary" goals/targets that specifically link climate change and sustainable energy objectives. At the intergovernmental level, these two global challenges – energy for sustainable development and global climate change – tend to be viewed and treated as distinct and separate issues with few, sustained attempts at policy coordination and synergy. The absence of a programmatic nexus between climate change and energy for sustainable development at the global level may be directly traced to the fact that intergovernmental negotiations on climate change have remained distinct from non-binding intergovernmental negotiations on energy for sustainable development within the context of the CSD. Although the central role of energy is widely acknowledged in the framing of the global climate change problem, global negotiations and dialogue on climate change under the aegis of the UNFCCC and its Kyoto Protocol phases have remained separate and distinct from non-binding global negotiations and dialogue on energy for sustainable development within the context of the CSD.

AGECC/SE4All initiative Following the failure of CSD-15 to arrive at a consensus-based decision on energy for sustainable development-related targets, the issue was given a global focus in 2010. The possibility of targets related to energy access and energy efficiency was given increased prominence with the 2010 launch of the UN Secretary-General's AGECC, the previously cited Energy for All global conference convened by Norway in 2011, and the subsequent UN declaration of 2012 as the International Year of Sustainable Energy for All. The AGECC represents an interesting institutional outcome because, on the one hand, it is the first UN-led initiative that was expressly "convened to address the dual challenges of meeting the world's energy needs for development while contributing to a reduction in GHGs" (AGECC, 2010, p. 3), while on the other hand, it is not an direct institutional outcome that emanates from an intergovernmental negotiation as it was convened at the behest of the UN Secretary-General as his advisory group. Comprising representatives from the private sector, the UN agencies, and research institutions, the AGECC was chaired by Kandeh Yumkella, Chair of UN-Energy and Director-General of the UN Industrial Development Organization.

In its official report, the AGECC called on the UN system and its member states to commit themselves to two complementary goals:

- Ensure universal access to modern energy services by 2030;
- Reduce global energy intensity by 40% by 2030.

Recognizing the challenges associated with defining global energy intensity, the AGECC report accepted energy efficiency as a proxy, as evidenced by its description of the actions required to meet the twin goals. Leaving aside the issue of mitigation targets, energy efficiency targets, particularly in the form of end-use efficiency, which is widely recognized to be the largest contributor to CO_2 emission reductions, are not specifically addressed within the current climate change framework (IEA, 2009, p. 46). But the issue of energy access is more complicated because, in spite of having energy access as one of its stated goals, the AGECC report also clearly pointed out that one of the challenges associated with energy access is that there is no definite global consensus on what the term "energy access" means. Derived from the IEA, AGECC lists three incremental levels of access to energy services:

- Level 1: Basic human needs
- Level 2: Productive uses
- Level 3: Modern society needs

The AGECC report defined universal energy access as "access to clean, reliable and affordable energy services for cooking, heating, lighting, communication and productive uses", in other words levels 1 and 2 (AGECC, 2010, p. 13).

The series of actions recommended by the AGECC towards achieving these twin goals were to launch a global campaign in support of "energy for sustainable development", which would be focused on improving energy access, enhancing energy efficiency, as well as raising awareness about the role of clean energy in addressing the MDGs; prioritize goals in countries through the adoption of appropriate national strategies; call on the international community to make available innovative financing mechanisms and climate finance; and emphasize and encourage private sector participation in achieving the goals (AGECC, 2010, p. 9).

In describing the financing needs for achieving the AGECC goals, the report estimated a need for $15 billion in grant financing mainly to cover capital investments and capacity-building in LDCs, and an additional $20–25 billion of loan capital for governments and private sector, an average of $30–35 billion of capital for energy efficiency in low-income countries and $140–170 billion for middle-income countries. But, most interestingly, the AGECC report noted that to support investment in energy access and energy efficiency, climate finance could be mobilized through funds being made available from the so-called $30 billion – Fast Track Funding – committed in COP-15 in Copenhagen. Also, conveniently, the AGECC recommended that in the medium to long term, another advisory group set up by the UN Secretary-General – the High-Level Advisory Group on Climate Change Financing – "could make it a priority to address the financing needs for energy efficiency and low-carbon energy access investments" (AGECC, 2010, p. 11).

By 2011, the AGECC had been subsumed by a newly launched initiative of the UN Secretary-General – the SE4All. Defined as a multi-stakeholder partnership among governments, the private sector, and civil society, it was launched by the UN Secretary-General in 2011, with three interlinked objectives to be achieved by 2030:

1. Ensure universal access to modern energy services.
2. Double the global rate of improvement in energy efficiency.
3. Double the share of renewable energy in the global energy mix.

According to the SE4All initiative website (http://www.se4all.org/about-us/), the three objectives reinforced each other in important ways, and achieving the three objectives together would allow for the maximization of development benefits and help to stabilize climate change over the long run. Currently, SE4All is a joint initiative of the UN and World Bank and is chaired by the UN Secretary-General and the President of the World Bank. SE4All now comprises an extensive advisory board, a separate executive committee, and a small global facilitation team and is working to launch a global practice network for energy access. Currently, there is an additional goal of renewable energy added to the twin goals identified previously by AGECC.

Global Sustainability Panel report In addition to the SE4All initiative, the UN also created a Global Sustainability Panel (GSP) in 2010, as an initiative of the Secretary-General. Co-chaired by President Tarja Halonen of Finland and President Jacob Zuma of South Africa, and comprising 20 members including Gro Harlem Brundtland, the GSP was to be a forum for high-level political participation that would lend weight and provide high-level inputs in the form of final report. According to the GSP website, a series of four GSP documents were prepared after periodic meetings of the GSP (New York, September 2010; Cape Town, February 2011, Helsinki, May 2011; and New York, September 2011), and in addition to the GSP there were "Sherpas", or experts, who also met and produced a series of different report (New York, October 2010; Braunwald, January 2011; Madrid, April 2011; New York, June 2011).

A background paper was prepared for the consideration of the GSP at its first meeting, entitled *Sustainable Development: From Brundtland to Rio 2012*. Although there was ample evidence regarding the linkages between lack of energy access and poverty reduction, and the AGECC had emphasized the role of energy services in driving sustainable development, this paper contained only one reference on access to energy in a section identifying progress on sustainable development and this reference downplayed the significance of the energy access problem (Drexhage and Murphy, 2010, p. 14).

The final report of the GSP, entitled *Resilient People, Resilient Planet: A Future Worth Choosing*, was launched in 2012 in Addis Ababa and also in various world capitals. The GSP report includes six sections:

- Panel's vision
- Progress towards sustainable development
- Empowering people to make sustainable choices
- Working towards a sustainable economy
- Strengthening institutional governance
- Conclusion: a call for action.

The report identifies a series of recommendations and, amongst the first, is the critical need to "embrace a new nexus between food, water and energy rather than treating them in different "silos" because all three "need to be fully integrated and not treated separately" in order to deal with the global food security crisis and, accordingly, a second green revolution would need to be embraced – an ever-green revolution for sustainability (UN Secretary-General's Global Sustainability Panel [GSP], 2012, p. 13).

The report points out that "1.3 billion people globally lack access to reliable electricity, while 2.7 billion people still rely on traditional biomass use for their cooking needs" and cited the IEA estimation that universal access to modern energy services by 2030 could be achieved at less than 3% of the total energy investment required by 2030, with a modest impact on total energy demand and carbon dioxide emissions (UN Secretary-General's GSP, 2012, p. 18). The report also references the issue of energy access in terms of the SE4All initiative (p. 43). It further pointed out that SE4All initiative provided proposals for what should be included in terms of SDGs and that energy issues "illustrate the cross-cutting challenge of sustainable development" in that "there is a social dimension to universal access to energy, an economic aspect to issues of affordability and energy efficiency, and an environmental side to emissions reduction – and all three are closely interlinked" (p. 73).

Although the food–water–energy nexus had been previously focused on by the WEHAB initiative, and by default was not a "new nexus", the GSP was clear about the scope of the energy access challenge. This clarity was not evident in terms of the GSP's focus on climate change. With regard to climate change, Dodds *et al.* (2014) noted that the GSP had a problem from the start identifying what the panel should address and its focus "became broader as the realization grew that climate change cannot be solved by itself … the final terms of reference of the GSP had more of a focus on climate change than the final report which really had very little" (p. 55). An examination of the terms of reference for the GSP reveals that it was intended to have a "special focus on climate change as sustainable development challenge", but what is even more relevant from the perspective of this chapter was that the GSP was expected to focus " in addition to climate change" on "other challenges which will be used to develop and test the new vision for sustainable development [which] may include food, water and energy security, as well as poverty reduction" (UN Secretary-General's GSP, 2012, p. 90).

An examination of the final GSP report reveals a paucity of concrete references and recommendations with regard to addressing climate change in relation to energy and poverty reduction. The report pointed out that "climate change is a risk to all countries and individuals", but noted that the "risks are particularly severe for the world's poorest"; however, it did not specifically identify the challenge of addressing poverty reduction in terms of climate change, instead listing "peace and security, and fairness in the distribution of responsibility and risk" as new challenges that emerge as a consequence of climate change (UN Secretary-General's GSP, 2012, p. 20) In the section dealing with empowering people to make sustainable choices, the report is emphatic in stating that those living in poverty "are also among the most vulnerable to the impacts of climate change, resource scarcity and environmental degradation" (p. 29), but there are no specific programmatic or policy linkages drawn to poverty reduction measures and climate change. And while there is a reference to the role of microfinance in reducing the vulnerability of the poor to the impacts of poverty and climate change (p. 61), the report's final recommendations tend to focus more on the role of macro green growth strategies that can respond to climate change by reducing carbon emissions rather than a specific focus on linking poverty reduction and climate change concerns, The focus is not on issues like short-lived climate pollutants and reducing indoor air pollution caused by lack of access to clean energy services, but on the role of climate-friendly technologies in dealing with global climate change in developing a green economy (UN Secretary-General's GSP, 2012, p. 82) and the macro linkages among energy, climate change, biodiversity and green growth (p. 85).

The issue of energy access for the poor was also focused on within the context of the UNGA. Resolution 65/151 of the UNGA recognized that:

> ...access to modern, affordable energy services in developing countries is essential for the achievement of the internationally agreed development goals, including the Millennium Development Goals, and sustainable development, which would help reduce poverty and to improve the conditions and standard of living for the majority of the world's population (UNGA, 2011).

The Resolution also designated 2012 as the International Year of Sustainable Energy for All. Whether this explicit recognition of energy access for the poor was adequately promoted in the 2012 Rio+20 Summit outcome is an important issue that will be considered in the following section.

Rio+20 Summit's *Future We Want* (FWW)   While initiatives like SE4All and the GSP were under way in 2010–2011, governments were gearing up to undertake the preparatory negotiations for the Rio+20 Summit. Like previous UN global conferences, the 2012 UN Conference on Sustainable Development (UNCSD), which is commonly referred to as the Rio+20 Summit (because it was convened 20 years after the 1992 historic Rio Earth Summit), included an extensive negotiating process of what is referred to in the UN as "informal informals", intersessional meetings, regional preparatory meetings and more formal preparatory committees, which were co-chaired by Ambassadors Kim Sook of South Korea and John Ashe of Antigua and Barbuda.

There was a first preparatory committee meeting held in May 2010, a second meeting held in March 2011, and a series of regional preparatory meetings following which a "Zero Draft" was prepared as the basis for negotiations. The release of this draft started a series of intense and arduous negotiations – informal informals sessions in March, end of April–May and from the end of May into June where delegates undertook long discussions and proposed a series of new textual additions which resulted in the draft FWW document comprising over 200 pages which were then being streamlined to 80 pages (*Earth Negotiations Bulletin* [ENB], 2012a). The third and last phase of the preparatory process began on June 13, and included as per established UN process, a literal line-by-line reading and negotiations which officially concluded at the closing of the third preparatory committee meeting on June 15, 2012 in Rio. The 1st plenary meeting of the UNCSD elected Ambassador Ashe (Antigua and Barbuda) as Chair of the Main Committee of the Conference, but the negotiations on the final outcome document were successfully concluded before the official start of the Conference on June 20, 2012. Unlike the down-to-the-wire global climate change negotiations for COPs that have extended a day or more past the official conclusion date, UNCSD was gaveled to end with an main outcome document – FWW – in time because there was an opportunity to have pre-conference intensive negotiations.

The UNCSD's principal focus was on two themes: a green economy in the context of sustainable development poverty eradication; and the institutional framework for sustainable development. According the UN, in the preparatory process, seven areas were highlighted as priority attention and included decent jobs, energy, sustainable cities, food security and sustainable agriculture, water, oceans and disaster readiness (see the UNCSD website, About Rio+20, www.uncsd2012.org/about.html). According to the ENB, 191 UN member states and observers were represented at Rio+20, including 79 heads of state or government, and over 44,000 official badges for attendance were given out. The ENB stated that voluntary commitments for actions to implement the conference's goals were encouraged at Rio+20, which lend further credence to the growth of partnership-related action in sustainable development and these included: a partnership between the US and African nations, with $20 million in funding announced by US Secretary of State Hillary Clinton, $6 million to UNEP's fund targeting developing countries, and $10 million towards climate change challenges in Africa, LDCs, and SIDS by pledge from President Rousseff, a similar pledge offered earlier in the meeting by Chinese Premier Wen Jiabao, the mobilization of €400 million to support sustainable energy projects announced by J.M. Barroso, President, European Commission, funding for a 3-year program of disaster risk reduction announced by Japan, and $175 billion over the next 10 years to support the creation of sustainable transport systems pledged by eight multilateral development banks (ENB, 2012b). The ENB analysis points out that broader partnership actions related to sustainable energy issues in general received considerable pledges of financial support, indicating that the issue of energy for sustainable development offers an exciting area for ongoing and future multilateral and civil society-driven action. The issue of non-UN led partnerships on energy access and energy and climate change are discussed at length in the following chapter.

The FWW was officially adopted by the UNGA Resolution 66/288 on July 27, 2012 (UNGA, 2012a). The FWW is a 53-page document comprising 283 paragraphs in all. The bulk of the FWW is contained in Section V of the FWW entitled, "Framework for action and follow up" which is subdivided into two sections: thematic areas and cross-sectoral issues; and sustainable development goals. The first section contains a vast array of thematic areas and cross-sectoral issues, including poverty eradication, energy, and climate change, which are all considered in distinct and different paragraphs. Box 3.6 provides a list of the various areas and issues that are listed in exactly the order they appear in the FWW and include the number of paragraphs focused on each issue listed in parentheses.

The FWW contains 29 references to "energy" – 24 of the total references to energy are contained in five contiguous paragraphs (para. 125–129) in the FWW's sub-section devoted to energy. What is particularly striking is that FWW included two separate paragraphs that were clear about the urgency of the energy access challenge. The introductory paragraph 125 recognized:

> …the critical role that energy plays in the development process, as access to sustainable modern energy services contributes to poverty eradication, saves lives, improves health and helps to provide for basic human need (UNGA, 2012a).

It stressed that these energy services were essential to social inclusion and gender equality and that energy was key to production, and it went even further by using the word "commit" to specify the supporting of energy access for the 1.4 billion people who currently lack access (UNGA, 2012a, p. 24). The next paragraph (para. 126) reiterated the energy access concern and stated:

> We emphasize the need to address the challenge of access to sustainable modern energy services for all, in particular for the poor, who are unable to afford these services even when they are available. We emphasize the need to take further action to improve this situation, including by mobilizing adequate financial resources, so as to provide these services in a reliable, affordable, economically viable and socially and environmentally acceptable manner in developing countries (UNGA, 2012a, p. 24).

The FWW however contains just one reference on SE4All contained in paragraph 129:

> We note the launching of the initiative by the Secretary-General on Sustainable Energy for All, which focuses on access to energy, energy efficiency and renewable energies. We are all determined to act to make sustainable energy for all a reality and, through this, help to eradicate poverty and lead to sustainable development and global prosperity. We recognize that the activities of countries in broader energy-related matters are of great importance and are prioritized according to their specific challenges, capacities and circumstances, including their energy mix (UNGA, 2012a, p. 25).

To be clear, in the lengthy process of negotiations, which included line-by-line readings and negotiations of the FWW text, there were stated concerns related to the SE4All initiative expressed by member states. Within the context of the

Box 3.6   *Thematic areas and cross-sectoral issues identified in the framework for action and follow-up of the* Future We Want *(FWW)*

- Poverty eradication (para. 105–107)
- Food security, nutrition and sustainable agriculture (para. 108–118)
- Water and sanitation (para. 119–124)
- Energy (para. 125–129)
- Sustainable tourism (para. 130–131)
- Sustainable transport (para. 132–133)
- Sustainable cities and human settlements (para. 134–137)
- Health and population (para. 138–146)
- Full productive employment, decent work for all and social protection (para. 147–157)
- Oceans and seas (para. 158–177)
- Small island developing states (para. 178–180)
- Least developing countries (para. 181)
- Land locked developing countries (para. 182)
- Africa (para. 183–184)
- Regional efforts (para. 185)
- Disaster risk reduction (para. 186–189)
- Climate change (para. 190–192)
- Forests (para. 193–196)
- Biodiversity (para. 197–204)
- Desertification, land degradation and drought (para. 205–209)
- Mountains (para. 210–212)
- Chemicals and waste (para. 213–223)
- Sustainable consumption and production (para. 224–226)
- Mining (para. 227–228)
- Education (para. 229–235)
- Gender equality (para. 236–214)

Source: UNGA (2012a).

UN, the use of the word "note" in connection with the SE4All initiative was an attempt to salvage this reference from being deleted, which was what the largest group of developing countries – the Group of 77 and China – had called for in the negotiating phase. The reference to the word "note" should be seen in comparison to the stronger word "recognize" contained in paragraphs 125–128 dealing with the broader role of energy in development and the role of energy efficiency, respectively. It should also be seen in comparison to the even more emphatic word "emphasize", which is contained in paragraph 126, focused expressly on the need to address the challenge of access to sustainable modern energy services for all; and the more purposeful word "reaffirm" contained in paragraph 127, which focused on support for the implementation of national and subnational policies and programs relating to the appropriate energy mix to meet developmental needs (UNGA, 2012a, pp. 24–25).

Despite the global launch of the SE4All initiative, and the inclusion of SE4All goals in the GSP final report, which was supposed to provide a direct input to the Rio+20 outcome document, the SE4All initiative did not get the prominence or the focus that was anticipated in the FWW.

The SE4All initiative received a lackluster global response in the FWW. Additional programmatic references and linkages could have been specified by FWW in terms of concrete energy access for the poor goals and targets. There was no global consensus to formally adopt the SE4All goals and there was no mention of the energy targets and goals in the FWW. Confirming this analysis, Dodds *et al.* (2014) noted that "there was reticence in the reaction by member states to the initiative" and "it was not endorsed or adopted by any of the regional or preparatory meetings for Rio+20 (except for the one in Barbados), or in the negotiations prior to the conference" (pp. 59–60).

The issue of climate change was referenced 22 times in the FWW document, whilst the issue of poverty eradication is referenced 36 times. As discussed earlier, in addition to energy, the FWW contained a large list of "thematic areas and cross sectoral areas" in section V, entitled "Framework for action and follow up". "Poverty eradication", which was listed as the first of such areas, was the focus of three paragraphs, and "Climate change", which was listed much further along, was the focus of three paragraphs. Some of the other thematic area issues included: food security (focus of 11 paragraphs), water and sanitation (focus of siz paragraphs), sustainable cities and human settlements (focus of four paragraphs), oceans and seas (focus of 20 paragraphs), LDCs (focus of one paragraph), Africa (focus of two paragraphs) and biodiversity (focus of eight paragraphs). But it is not clear why the non-alphabetical ordering of the thematic areas and cross-sectoral issues and different issues are given varying levels of textual attention as evidenced earlier (UNGA, 2012a).

The climate change challenge was not highlighted in the initial segment of the FWW. The first references to climate change are in paragraph 17 in relation to supporting the Rio Convention and the UNFCCC; and later in paragraph 25, when the FWW acknowledged that "climate change is a cross-cutting and persistent crisis" and expressed:

> ...concern that the scale and gravity of the negative impacts of climate change affect all countries and undermine the ability of all countries, in particular, developing countries, to achieve sustainable development and the Millennium Development Goals, and threaten the viability and survival of nations (UNGA, 2012a, p. 6).

Here, the need for urgent and ambitious action to combat climate change in accordance with the principles and provision of the UNFCCC was highlighted. There is a three-paragraph focus on climate change in the FWW. In paragraph 190, climate change is reaffirmed as "one of the greatest challenges of our time" and here the FWW expresses "profound alarm that emissions of greenhouse gases continue to rise globally". Also in this paragraph, deep concern is expressed for the vulnerabilities of all countries and particularly developing ones to climate change, and here there is a mention of the adverse impacts of climate change "further threatening food security and efforts to eradicate poverty", which leads to adaptation to climate change being accorded as "an immediate and urgent global priority" (UNGA, 2012a, p. 37).

Paragraphs 191 and 192 are two of the remaining three paragraphs focused on climate change, which is where the bulk of the references to climate change are contained. What is important to note is that paragraph 191 was explicit in calling for the "widest possible cooperation by all countries and their participation in an effective and appropriate international response, with a view to accelerating the reduction of global greenhouse gas emissions" (UNGA, 2012a, p. 37). As such, the FWW call for broader-scale participation in addressing the climate change mitigation echoed the ongoing trend in the UNFCCC negotiations, but there was a general lack of policy guidance on energy, and in particularly energy access for the poor and its implications for addressing climate change vulnerabilities and mitigation.

The FWW section dealing with SDGs is crucial in terms of any future resolution of energy for sustainable development and climate change goals, including energy access for the poor goals and targets. The FWW reaffirmed the importance of achieving the MDGs and pointed out that the SDGs should not divert focus or effort from the MDGs (UNGA, 2012a, p. 47). But importantly, in paragraph 247, the FWW explicitly recognized that any future SDGs:

> ...should be action oriented, concise and easy to communicate, limited in number, aspirational, global in nature and universally applicable to all countries, while taking into account different national realities, capacities and levels of development and respecting national policies and priorities (UNGA, 2012a, p. 47)

The FWW called for establishing "an inclusive and transparent intergovernmental process" for developing the SDGs that were to be agreed by the UNGA and mandated the creation of an inter-governmental Open Working Group "no later than the opening of the 67th session of the UNGA" that would "submit a report to the 68th session of the General Assembly containing a proposal for SDGs for consideration and appropriate action" (UNGA, 2012a, p. 47). The FWW is clear that "the process leading to the SDGs needs to be coordinated and coherent with the processes considering the post 2015 development agenda" and that the "initial input to the work of the Open Working Group will be provided by the UN Secretary General in consultation with national governments" (UNGA, 2012b, p. 47).

3.5 Conclusion: Implications for the energy–poverty–climate change nexus in the UN's post-2015 development agenda

In terms of responding to energy with the UN context, Najam and Cleveland (2005) state that energy has received increasing prominence in successive environmental summits, starting from 1972 and going through to 2002. But based on a close examination of the 40-year record of UN conference outcomes, this chapter demonstrates quite the opposite. Energy for sustainable development has been dealt with in a scattershot manner over the course of various key UN global outcomes. There is little to no evidence in the key UN globally agreed outcomes on sustainable development that can corroborate increasing focus on energy and/or detailed attention paid to the linkages among energy, poverty reduction, and climate change. The only exception to this was the WEHAB's *A Framework for Action on Energy* background paper, which outlined indicative targets and examples of activities in as many 12 action areas related to energy for sustainable development and energy access for the poor.

The detailed examination of a series of UN-led global conferences and initiatives related to environment and development and sustainable development, from the 1972 UNCHE to the 2012 Rio+20 Summit, reveal a disjointed approach and an unclear trajectory in terms of how energy issues and, in particular, energy access for the poor have been dealt with over the course of the past 40 years. Despite the recommendations of numerous UN and global reports on the importance of energy for sustainable development and poverty reduction, the review of all the major UN-led global outcomes demonstrates that:

- Detailed references to energy access for the poor that have any programmatic implications are largely absent, and there remain no specific targets or goals related to this issue to date.
- Energy for sustainable development, including energy access objectives, and climate change objectives have quite consistently been treated as separate and distinct negotiated outcomes, with attendant implications for separate follow-up.

Notwithstanding the convening of global conferences and high-level expert panels, concrete recommendations and guidelines focused on the energy–poverty–climate change nexus are largely missing in the key agreed global sustainable development outcomes examined. References linking energy access for the poor and climate change have not been specified in almost all the global outcomes discussed earlier.

The 1972 UNCHE set the threshold in terms of formulating a host of environmental governance ideas, many of which had a lasting impact on global environmental governance. Although it can be seen as referencing the first intergovernmental negotiated recommendation for having networks and guidelines for monitoring emissions of gases related to energy services, its recommendations relating to energy were more focused on data and information gathering, and there was no explicit linkage between energy services and poverty reduction and there was no mention of the issue of climate change. The WCED provided the historic definition of sustainable development, which is a concept that has had great resonance and longevity in terms of being referenced both as a policy tool and within UN-led global negotiations. The WCED also provided a much more detailed understanding and set of recommendations related to the linkages between energy and climate change. It highlighted the role of energy in promoting human development and meeting basic needs, and was prescient in outlining the climate change challenge, including the notion of the precautionary principle, as well as health and sustainability concerns over the dependence on traditional biomass in developing countries – both of which remain huge and urgent development priorities.

Agenda 21, with its 40 voluminous chapters, did not really build on the trajectory and momentum of the WCED in terms of the outlining the role of energy in human development and specifying the linkages between energy and climate. For instance, unlike the WCED, which had a specific chapter on energy, references to energy were much more scattered in Agenda 21. The lack of explicit linkages between energy and poverty reduction and/or human well-being improvement appears to be largely missing in Agenda 21 and may be explained by the fact that the principal focus of Agenda 21 was more environmental than socio-economic. The fact of the matter is that the 1992 Rio Summit was the forum for the historic UNFCCC and, accordingly, climate change issues were seen as being dealt within that framework convention and separate from the broader global negotiations on sustainable development. This point needs to be underscored, because, by and large, the bifurcation has been held for the past two decades and climate change within the UN context has been seen as an issue falling within the purview of the UNFCCC, which has legally binding implications. The diminished references to the urgency of climate change, which is dealt with in the context of the UNFCCC, and the more scattershot approach to energy are evident in Agenda 21.

There is an absence of any detailed policy and programmatic nexus between climate change and energy for sustainable development, including energy access for the poor at the global level, which may be directly traced to the fact that global negotiations on climate change within the UNFCCC have remained distinct from intergovernmental discussions on sustainable development within the CSD. Since early 2000, issues such as accessibility of energy, energy efficiency, renewable energy, advanced fossil fuel technologies, rural energy, and energy and transport have been the focus of attention within the context of the energy for sustainable development negotiations, but they have not been the focus of policy attention within the context of the global climate change negotiations.

Energy for sustainable development in general, and energy access for the poor are well researched issues and have received particular attention from a host of UN and global agencies from the late 1990s onwards. But rather than steadily gaining prominence and global attention in terms of UN-led conferences and outcomes, energy issues in general and the linkages between energy access and poverty reduction and climate change seem to be largely absent or puzzlingly disconnected in the agreed global outcomes and initiatives examined in this chapter. Even though there was ample evidence linking energy services to a host of development challenges, including poverty reduction, maternal and infant mortality and ill-health, gender inequality, water and sanitation, food security and others, neither the Millennium Declaration nor the MDGs included a concrete reference to the vital role of increasing access to cost-effective and sustainable energy services. Notwithstanding the findings of several influential UN-sponsored studies that highlighted the importance of energy in reducing poverty and improving the health and mortality of poor women and children in developing countries from 2000 onwards, the 2000 Millennium Declaration, its attendant MDGs, and the JPOI had no energy access for the poor target or goal, and no such target or goal was accepted or endorsed by the Rio+20 Summit.

The JPOI continued the trend of having a more diffused approach to energy, and again it did not contain any specific chapter focused on energy, nor were any concrete energy access for the poor targets or recommendations adopted, despite the JPOI's emphasis on linking global poverty reduction efforts with broader SDGs. What was novel about the JPOI was the role of partnerships, which is an area where a considerable amount of non-UN-led stakeholder action is happening in terms of the energy–poverty–climate change nexus, as the following chapter will discuss.

Meanwhile in the context of the UNGA having declared 2012 the International Year of Sustainable Energy for All, the 2012 FWW outcome contained a rather marked paucity of concrete recommendations on how to actually deliver on the goal of energy access for the poor. The issues of climate change and energy were dealt with in separate sections of the document, and received less attention than other thematic issues. In spite of the UN and World Bank-sponsored SE4All initiative – which expressly focused on increasing energy access as one of three globally publicized goals – and the high-profile

GSP, which endorsed the goals of the SE4All, the FWW document did not endorse or support any goals, but ended up only taking "note" of the SE4All initiative. So the question is, what will happen to the issue of energy access for the poor in the post-2015 UN development agenda?

Even though there was no resounding affirmation in the FWW, there has been a flurry of new panels and ramping up of the SE4All initiative in the post-2012 phase. The UN Secretary-General established the High-Level Panel (HLP) in July 2012, right after the Rio+20 conference comprising 26 members. The HLP was tasked to provide "bold yet practical thinking for the post-2015 development agenda" and its report, released in May 2013, is entitled, *A New Global Partnership: Eradicate Poverty and Transform Economies through Sustainable Development*. The report highlights the need for a single agenda that brings together social, economic, and environmental issues, and for a universal agenda that is relevant to, and actionable by, all countries. It clearly states that "all countries should have access to cost-effective, clean and sustainable energy" (UN, 2013, p. 15) and in a box listing what the world would have by 2030, it lists 1.2 billion more people connected to electricity (UN, 2013, p. 19).

The report, which keeps poverty eradication at its core, argues that the post-2015 development agenda needs to be driven by five big transformative shifts: leave no one behind; put sustainable development at the core; transform economies for jobs and inclusive growth; build peace and effective, open and accountable institutions; and forge a new global partnership. While these broad declarative issues are not easy to translate into practical action, the report's most definitive contribution is its comprehensive list of 12 universal goals – which form a sort of expanded list of MDGs – including a specific goal on sustainable energy. Goal 7, entitled "secure sustainable energy", has four distinct targets:

- Double the share of renewable energy in the global energy mix.
- Ensure universal access to modern energy services.
- Double the global rate of improvement in energy efficiency in buildings, industry, agriculture. and transport.
- Phase out inefficient fossil fuel subsidies that encourage wasteful consumption.

To be clear, the energy-related recommendations of the HLP are not the result of intergovernmental negotiations and can be contrasted with the development of any energy-related target or goal developed within the context of the intergovernmental negotiation on the SDGs.

The FWW mandated a basic timeline and process for developing the SDGs but it left it up to UN member states to figure out the exact modalities for the operation of the Open Working Group (OWG) which would propose the anticipated SDGs.

After considerable debate among UN member states, the OWG was established on January 22, 2013, close to 6 months after the FWW was adopted. UN member states decided to use an innovative, constituency-based system of representation that is new to limited membership bodies of the General Assembly and meant that each seat in the Group is shared by one to four member states. These country teams have decided internally how they will be represented in the OWG's various meetings.

Based on the FWW's recommendation, in December 2012, the UN Secretary-General prepared an input report that was to inform the work of the OWG. This input report is a synthesis of 63 member state responses, including a joint response of the European Union to a questionnaire sent out to member states regarding the SDGs, and represents roughly one-third of the United Nations membership. Figure 3.3, excerpted from this input report, reveals that energy was accorded more priority and ranked third, whilst climate change is ranked eighth in terms of frequency of responses in a long list of issues. The report references "energy" 20 times, and while the term "energy access" is not mentioned, the term "access to energy" is referenced twice. The first reference is contained in paragraph 21 which points out that "many respondents cited" the SE4All initiative and its targets and also "a target on access to energy" (UNGA, 2012b, p. 4). The second reference is clearer but vague in terms of implementation and delivery in that the report states that "for a developing country, access to energy remains important and could be supported by developed countries in the context of development cooperation" (UNGA, 2012b, p. 9). Meanwhile, climate change is referenced six times in the report but the linkages between energy and climate change are not mentioned.

From March to July 2013, the OWG met 13 times and conducted wide-ranging discussions on the SDGs, which included extensive consultations and inputs from civil society stakeholders. At its final, 13th session in July 2013, these civil society groups (also known as major groups and stakeholders) provided a compilation of their final amendments to goals and targets under consideration (UN SD Knowledge Platform, 2014). Unlike the MDGs, which were the result of a largely internal UN-driven exercise, the process for formulating and securing agreement on the SDGs has been much broader and more inclusive.

The final session of the OWG agreed on a set of 17 SDGs on July 19, 2013. The Report of the OWG on Sustainable Development, which was agreed to by the UNGA, lists the 17 SDGs, and includes a specific SDG on energy access (goal 7) and a separate stand-alone SDG on climate change (goal 13). The OWG report highlighted the overwhelming importance to be accorded to poverty eradication within the context of the SDGS, "poverty eradication is the greatest global challenge facing the world today and an indispensable requirement for sustainable development" (UNGA, 2014, p. 6), and also "underscored"

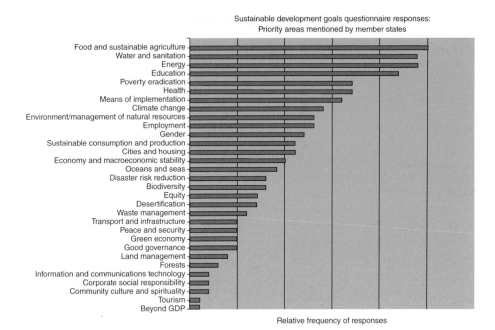

Figure 3.3 Sustainable development goals responses: priority areas ordered by frequency of responses by member states (source: UNGA, 2012b).

that "the global nature of climate change calls for the widest possible cooperation by all countries and their participation in an effective and appropriate international response, with a view to accelerating the reduction of global greenhouse gas emissions" (UNGA, 2014, p. 8). The report was also clear that the SDGs "depend on a global partnership for sustainable development with the active engagement of Governments, as well as civil society, the private sector and the United Nations system" and noted that a "robust mechanism to review implementation will be essential for the success of the goals" (UNGA, 2014, p. 9). The General Assembly, the Economic and Social Council, and the high-level political forum will play a key role in this regard. Given the FWW's recommendation that the SDGs be universal in scope, many of the goals are broad declarations of intent. Table 3.1 provides a list of the 17 SDGs and the number of proposed targets associated with each SDG. There is a staggering total list of 169 targets associated with the SDGs, which makes for an extraordinarily complicated implementation phase of any post-2015 development agenda charged with monitoring and evaluating progress made towards achieving individual goals and their associated targets.

The final report of the OWG on SDGs reveals a unique dichotomy between the one SDG on climate change and the remaining 16 (UNGA, 2014b). What is clear from the final outcome of the OWG on SDGs is that goal 13 on climate

Table 3.1 List of sustainable development goals and their associated targets

Sustainable development goals	Number of associated targets
Goal 1 – End poverty in all its forms everywhere	7
Goal 2 – End hunger, achieve food security and improved nutrition, and promote sustainable agriculture	8
Goal 3 – Ensure healthy lives and promote well-being for all at all ages	13
Goal 4 – Ensure inclusive and equitable quality education and promote life-long learning opportunities for all	10
Goal 5 – Achieve gender equality and empower all women and girls	9
Goal 6 – Ensure availability and sustainable management of water and sanitation for all	8
Goal 7 – Ensure access to affordable, reliable, sustainable, and modern energy for all	5
Goal 8 – Promote sustained, inclusive and sustainable economic growth, full and productive employment and decent work for all	12
Goal 9 – Build resilient infrastructure, promote inclusive and sustainable industrialization and foster innovation	8
Goal 10 – Reduce inequality within and among countries	10
Goal 11 – Make cities and human settlements inclusive, safe, resilient and sustainable	10
Goal 12 – Ensure sustainable consumption and production patterns	11
Goal 13 – Take urgent action to combat climate change and its impacts* (*Acknowledging that the UNFCCC is the primary international, intergovernmental forum for negotiating the global response to climate change)	5
Goal 14 – Conserve and sustainably use the oceans, seas and marine resources for sustainable development	10
Goal 15 – Protect, restore and promote sustainable use of terrestrial ecosystems, sustainably manage forests, combat desertification, and halt and reverse land degradation and halt biodiversity loss	12
Goal 16 – Promote peaceful and inclusive societies for sustainable development, provide access to justice for all and build effective, accountable and inclusive institutions at all levels	12
Goal 17 – Strengthen the means of implementation and revitalize the global partnership for sustainable development	19
Total = 17 goals	**Total =169 targets**

Source: UNGA (2014, pp. 10–24).

change will be treated differently from the others. Goal 13 clearly indicates that the locus of intergovernmental action on climate change does not lie with the UNGA but instead will be considered separately within the context of the UNFCCC's intergovernmental negotiations. There is a distinction drawn between the intergovernmental consideration of energy for SDGs and targets and the climate change goals and targets that will be arrived at separately under the aegis of the UNFCCC process. What this means is that although the FWW has mandated that the SDGs be agreed to by the UNGA, any future climate change-related goals and targets can, and will only, be agreed by climate change negotiators at COP-21 in December 2015. Accordingly, it is possible that the final global agreement on SDGs anticipated for September 2015 within the context of the UNGA will be pending any future SDG-related goals and targets on climate change based on whatever global agreement is achieved in December at COP-21 in Paris. So, in all likelihood, a default placeholder will have to be maintained for the SDGs on climate change and the UNGA will have to await the final agreed outcome SDG goal, which may or may not be forthcoming in terms of concrete associated targets.

What is also important to point out is that the specific targets associated with goal 7 (energy access for all) and goal 13 (climate change), which are listed in Box 3.7, demonstrate that there are no established linkages or references that point to the possibility of integrated action on the nexus between energy access for the poor and climate change. The targets associated with energy access for all mention 2030 as the deadline for ensuring the meeting of targets, but the targets associated with climate change are more generic and generalized and include no such deadlines. The fact of the matter is that any agreement secured on an SDG related to climate change and any associated targets arrived at within the UNGA are dependent on the level of ambition and consensus that can be achieved within the UNFCCC process. But what is truly significant about this reversal of dependence is that the UNGA is "the main deliberative, policy making and representative organ of the UN" (see the UN website: www.un.org/en/ga/about/index.shtml). Comprising all member states of the United Nations, it provides the most inclusive, forum for multilateral discussion of the full spectrum of international issues covered by the Charter of the UN. However, any final 2015 agreement on global climate change will be arrived at within the separate intergovernmental context of the UNFCCC process and would then be inserted into the agreed text on the SDGs.

From an intergovernmental perspective, the overall lack of global will to embrace targets and commitments in terms of both energy access and climate change remains a huge challenge for the UN as it embarks on formulating a universal post-2015 development. Given the past record of global outcomes, it is not immediately clear whether and how exactly the energy–poverty–climate change nexus will be dealt with in the final intergovernmental negotiations related to the adoption of the post-2015 development agenda, including the finally agreed SDGs. The silo approach to energy for sustainable development objectives, including energy access for the poor and global climate change, inhibits an integrated framework for sustainable development. Global climate change negotiations are distinct and separate from global negotiations on energy for sustainable development and this in turn has led to the puzzling absence of concrete linkages between energy access and climate change objectives. This

Box 3.7 Associated targets for proposed sustainable development goal 7 (energy access for all) and goal 13 (climate change)

Goal 7. Ensure access to affordable, reliable, sustainable and modern energy for all

7.1   By 2030, ensure universal access to affordable, reliable and modern energy services

7.2   By 2030, increase substantially the share of renewable energy in the global energy mix

7.3   By 2030, double the global rate of improvement in energy efficiency

7.a   By 2030, enhance international cooperation to facilitate access to clean energy research and technology, including renewable energy, energy efficiency and advanced and cleaner fossil-fuel technology, and promote investment in energy infrastructure and clean energy technology

7.b   By 2030, expand infrastructure and upgrade technology for supplying modern and sustainable energy services for all in developing countries, in particular least developed countries and small island developing states

Goal 13. Take urgent action to combat climate change and its impacts*

13.1   Strengthen resilience and adaptive capacity to climate-related hazards and natural disasters in all countries

13.2   Integrate climate change measures into national policies, strategies and planning

13.3   Improve education, awareness-raising and human and institutional capacity on climate change mitigation, adaptation, impact reduction and early warning

13.a   Implement the commitment undertaken by developed-country parties to the United Nations Framework Convention on Climate Change to a goal of mobilizing jointly $100 billion annually by 2020 from all sources to address the needs of developing countries in the context of meaningful mitigation actions and transparency on implementation and fully operationalize the Green Climate Fund through its capitalization as soon as possible

13.b   Promote mechanisms for raising capacity for effective climate change-related planning and management in least developed countries, including focusing on women, youth and local and marginalized communities

*Acknowledging that the United Nations Framework Convention on Climate Change is the primary international, intergovernmental forum for negotiating the global response to climate change.

Source: UNGA (2014, pp. 15, 19–20).

absence of a concrete nexus between climate change and energy for sustainable development at the level of intergovernmental negotiations impacts on the post-2012 climate change framework, because it limits practical scalable emissions reductions that derive directly from nationally driven energy sector needs and concerns. Having separate intergovernmental negotiations on two global policy challenges that inherently lend themselves to synergistic action does not make sense, inhibits innovation and cooperation, and is an unfortunate drain on a limited financial resources and technical capacities at all levels.

References

AGECC (2010) *Energy for a Sustainable Future: Summary Report and Recommendations*. New York: UN.

Caldwell, L. (1984*) International Environmental Policy: Emergence and Dimensions*. Durham, NC: Duke University Press.

Chasek, P. and Wagner, L. (eds) (2012) *The Road from Rio: Lessons Learned from Twenty Years of Multilateral Environmental Negotiations*. New York: RFF.

Cherian, A. (2012) Confronting a multitude of multilateral environmental agreements. In: Harris, F., ed., *Global Environmental Issues*. Oxford: Wiley-Blackwell Publications, pp. 39–61.

Department for International Development (2002) *Energy for the Poor*. UK: DFID.

Dodds, F., Laguna-Celis, J. and Thompson, L. (2014) *From Rio+20 to a New Development Agenda*. New York: Routledge.

Drexhage, J. and Murphy, D. (2010) *Sustainable Development: From Brundtland to Rio 2012. Background Paper prepared for the GSP*. New York: UN. http://www.un.org/wcm/webdav/site/climatechange/shared/gsp/docs/GSP1-6_Background%20on%20Sustainable%20Devt.pdf [accessed November 29, 2014].

ENB (2012a) United Nations Conference on Sustainable Development. *Earth Negotiations Bulletin*, 27(41), June 13, 2014. IISD.

ENB (2012b) Summary of the United Nations Conference on Sustainable Development. *Earth Negotiations Bulletin*, 27(51), June 25, 2012. IISD.

Flavin, C. and Lenssen N. (1994) *Power Surge*. New York: W.W. Norton

Goldemberg, J., Johansson, T., Reddy, A.K. and Williams, R.H. (1985) Basic needs and much more with one kilowatt per capita. *Ambio*, 14(4), 190–200.

Goldemberg, J., Johansson, T.B., Reddy, A.K.N. and Williams, R.H. (1988) *Energy for a Sustainable World*. New Delhi: Wiley Eastern Limited.

Haas, P., Levy, M. and Parson, E. (1992) Appraising the Earth Summit: How we should judge UNCED's success? *Environment*, 34(8), 26–33.

IEA (2009) *World Energy Outlook: 2009*. Paris: OECD/IEA.

IEA (2011) *Energy for All: Financing Access for the Poor*. Paris: OECD/IEA. http://www.iea.org/publications/freepublications/publication/weo2011_energy_for_all.pdf [accessed on February 27, 2014].

IEA (2013) *World Energy Outlook: 2013*. Paris: OECD/IEA.

IEA/UNDP/UNIDO (2010) *Energy Poverty: How to Make Modern Energy Access Universal*. Paris: OECD/IEA.

Johansson, T. and Goldemberg, J. (eds) (2002) *Energy for Sustainable Development: A Policy Agenda.* New York: UNDP.

Jordan, A. and Voisey, H. (1998) The 'Rio Process': The Politics and Substantive Outcomes of 'Earth Summit II' – Institutions for global environmental change. *Global Environmental Change,* 8(1), 93–97.

Klein, H. (2005) Understanding WSIS: An institutional analysis of the UN World Summit on the Information Society. *Information Technologies and International Development,* 1(3–4), 3–13.

Leach, G. (1992) The energy transition. *Energy Policy.* Vol. 20: 2. p.116-123.

Modi, V., S. McDade, D. Lallement, and Saghir, J. (2006) *Energy and the Millennium Development Goals.* New York: Energy Sector Management Assistance Programme, UNDP/UN Millennium Project/World Bank.

Munasinghe, M. (2002) The sustainomics trans-disciplinary meta-framework for making development more sustainable: applications to energy issues. *Int. J. of Sustainable Development,* Vol. 5: 1/ 2, p. 125-182

Najam, A. (2005) Developing countries and global environmental governance: from contestation to participation to engagement. *International Environmental Agreements,* 5(3), 303–321.

Najam, A. and Cleveland, C. (2005) Energy and sustainable development at global environmental summits: an evolving agenda. In: Hens, L. and Nath, Bhaskar, eds. *The World Summit on Sustainable Development.* Netherlands: Springer.

Nussbaumer, P., Bazilian, M. and Modi, V. (2012) Measuring energy poverty: Focusing on what matters. *Renewable and Sustainable Energy,* 16(1), 231–243.

Sokona, Y., Sarr, S., Wade, S. and Togola I., 2004, Energy Services for the poor in West Africa: Sub-regional "energy access" study of West Africa. Paper presented at the *Global Network on Energy for Sustainable Development.*

Sovacool, B. (2012) The political economy of energy poverty: A review of key challenges. *Energy for Sustainable Development,* 16(3), 272–282.

Srivastava, L., Goswami, A., Diljun, G.M. and Chaudhury, S. Energy access: revelations from energy consumption patterns in rural India. *Energy Policy,* 47(1), 11–20.

Srivastava, L. and Sokona, Y., eds. (2012) Universal access to energy: Getting the framework right. *Energy Policy,* 47(1), 1–87.

Strong, M. (1973) One year after Stockholm: An ecological approach to management. *Foreign Affairs,* 51, 690–707.

UN (1987) Report of the World Commission on Environment and Development. Available at *A/42/427. Our Common Future: Report of the World Commission on Environment and Development* Accessed on February 27, 2013.

UN (1992) Agenda 21. http://sustainabledevelopment.un.org/content/documents/Agenda21.pdf [accessed February 25, 2014].

UN (2001) *Comission on Sustainable Development: Report on the Ninth Session.* E/CN.17/2001/19. New York: UN.

UN (2002) *Johannesburg Declaration on Sustainable Development and Plan of Implementation on the World Summit on Sustainable Development.* New York: UNDESA. (WSSD report is also available as UN document A/CONF.199/20.)

UN (2003) *Commission on Sustainable Development: Report of the Eleventh Session.* (Annex to resolution for multi-year programme of work of CSD). E/CN.17/2003/6. New York: UN.

UN (2007) *Commission on Sustainable Development: Report on the Fifteenth Session*. E/CN.17/2007/15 New York: UN

UN (2013) *A New Global Partnership: Eradicate Poverty and Transform Economies through Sustainable Development*. New York: UN Publications.

UNDESA website. Major Conferences and Summits. Available at http://www.un. org/en/development/desa/what-we-do/conferences.html [accessed on February 25, 2014].

UNGA (1983) Process of preparation of the Environmental Perspective to the Year 2000 and Beyond. Resolution A/RES/38/161.New York: UN

UNGA (1987a) Environmental Perspective to the Year 2000 and Beyond. Resolution A/42/186. New York: UN.

UNGA (1987b) Report of the World Commission on Environment and Development A/RES/42/187. New York: UN

UNGA (1992) Report of the United Nations Conference on Environment and Development. A/CONF.151/26. New York: UN

UNGA (1997) Programme for the Further Implementation of Agenda 21 A/RES/ S-19/2 New York: UN

UNGA (2000) United Nations Millennium Declaration". Resolution A/RES/55/2. New York:UN

UNGA (2001) Road map towards the implementation of the United Nations Millennium Declaration (Report by the Secretary General). A/56/326. New York: UN.

UNGA (2011) International Year of Sustainable Energy for All. United Nations. A/RES/65/151. New York: UN

UNGA (2012a) *The Future We Want*. Resolution adopted by the General Assembly on 27 July 2012. A/RES/66/288. New York: UN.

UNGA (2012b) Initial input of the Secretary-General to the Open Working Group on Sustainable Development Goals. A/67/634. New York: UN. Available at http://www. un.org/ga/search/view_doc.asp?symbol=A/67/634&Lang=E Accessed on December 1, 2014.

UNGA (2014) Resolution adopted by the General Assembly on 10 September 2014: Report of the Open Working Group on Sustainable Development. A/Res/68/309. New York: UN

UNGA (2014) Report of the Open Working Group of the General Assembly on Sustainable Development Goals. A/69/970. New York: UN. Available at http://www. un.org/ga/search/view_doc.asp?symbol=A/68/970&Lang=E Accessed on December 16 , 2014.

UN Economic and Social Council (1999) Initiation of preparations for the ninth session of the Commission on issue related to the sectoral theme: energy. E/CN.17/1999/8. New York: UN

UNECOSOC (2001) *Report of the Secretary General: Energy and Transport*. E/CN.17/2001/PC/20 New York: UN

UN Millennium Project (2006) "About the MDGs". http://www.unmillenniumproject. org/goals/index.htm [accessed on Feb 16, 2014].

UN News Centre (2002a) "UN summit ends with action plan to boost efforts to protect environment, fight poverty", 4 September 2002. http://www.un.org/apps/news/

story.asp?NewsID=4617&Cr=johannesburg&Cr1=summit&Kw1=World+Summit &Kw2=&Kw3=#.Uw-zA9iUMUI [accessed on Feb 16, 2014]

UN News Centre (2002b) "UN summit 'major leap forward' for partnerships in global fight against poverty", 4 September 2002. http://www.un.org/apps/news/story.asp? NewsID=4615&Cr=johannesburg&Cr1=summit&Kw1=World+Summit&Kw2= &Kw3= [accessed May 11, 2015].

UN SD Knowledge Platform (2014) *Outcome Document: Introduction to the Proposal of the Open Working Group on Sustainable Development Goals.* http://sustainabledevelopment.un.org/content/documents/4518SDGs_ FINAL_Proposal%20of%20OWG_19%20July%20at%201320hrsver3.pdf [accessed September 15, 2014].

UN website. *Sustainable Development, Human Settlements and Energy.* http://www.un.org/en/development/progareas/dsd.shtml [accessed February 28, 2014].

UN website. UN Energy. http://www.un-energy.org/ Accessed on February 26, 2014.

UN website. "Rio+20 Conference". Available at http://www.uncsd2012.org/about.html. Accessed on February 27, 2014.

UN website. About the General Assembly. http://www.un.org/en/ga/about/index.shtml [accessed June 1, 2014].

UNDP/WHO (2009) *The Energy Access Situation in Developing Countries: A Review Focusing on the Least Developed Countries and Sub-Saharan Africa.* New York: UNDP.

UNDP/UNDESA/WEC. (2000) *World Energy Assessment: Energy and the Challenge of Sustainability.* New York: UNDP.

UN Secretary-General's Global Sustainability Panel (2012). Resilient People, Resilient Planet: A future worth choosing. New York: United Nations.

UN/WEHAB Working Group (2002) *A Framework for Action on Energy.* New York: UN. http://www.un.org/jsummit/html/documents/summit_docs/wehab_papers/wehab_ energy.pdf [accessed May 11, 2015].

WHO (2000) Addressing the links between indoor air pollution, household energy and human health: Based on WHO/USAID Global Consultation. Geneva: WHO.

WHO/UNDP (2009) *The Energy Access Situation in Developing Countries.* New York: UNDP.

4
Understanding the Acronym Soup of Voluntary Initiatives and Partnerships on Sustainable Development Within the UN Context

Locating Energy Access and Climate Change Voluntary Efforts

4.1 Delving into the acronym soup of voluntary initiatives for sustainable development at the UN

One of the most significant outcomes of the past decade has been the dramatic increase in the involvement of non-state actors first referred to in Agenda 21 (UN, 1992) as "major groups" and otherwise also known as civil society actors. Within the UN context, the involvement of the nine "major groups", defined as comprising business and industry, children and youth, farmers, indigenous peoples, local authorities, non-governmental organizations (NGOs), the scientific and technological community, women, and workers and trade unions, have been seen as key to ensuring the successful implementation of

Energy and Global Climate Change: Bridging the Sustainable Development Divide,
First Edition. Anilla Cherian.
© 2015 John Wiley & Sons, Ltd. Published 2015 by John Wiley & Sons, Ltd.

the 1992 Agenda 21, the 2002 Johannesburg Plan of Implementation (JPOI) and, more recently, the 2012 *Future We Want* (FWW; UNGA, 2012). The active engagement of a diverse range of civil society stakeholders in UN-led sustainable development fora, including the participation of all "major groups" in the formulation of the sustainable development goals (SDGs), is clear.

Unprecedented levels of civil society involvement have occurred on issues ranging from climate change to human rights, from biotechnology to land mines. The role of civil society participation has been cited as a significant factor in ensuring the efficacy of global environmental governance (Gemmill *et al.*, 2002). The need for multi-stakeholder participation at every stage of decision-making and implementation of sustainable development objectives has been clearly reaffirmed in UN global outcomes pertaining to sustainable development issues, and it is this broader context of increased civil society participation that has provided both the context and the impetus for the dynamic growth in voluntary efforts and initiatives relating to the gamut of sustainable development concerns. There has been upsurge of voluntary efforts, including partnerships for sustainable development (PSDs), as well as voluntary coalitions for action formed by key civil society actors and groups of governments in the past decade, and such efforts can be seen as portents of the future.

The concept of voluntary initiatives including PSDs appears to be tailor-made for the UN-led global quest for sustainable development and a better future for all, because it channels and builds on multi-stakeholder engagement and action. Voluntary initiatives are, after all, seen as promoting multi-stakeholder collaboration, which in turn has been seen as an essential element of inclusive and participatory sustainable development efforts at all levels. A review of the family of UN agencies reveals that almost every single UN agency and entity mandated to work on specific aspects of sustainable development at both the programmatic or policy levels has, and is currently engaged with, various forms of voluntary and/or partnership driven efforts in the fulfillment of their specific mandates. In his 2013 interview with *The Guardian* on the growing role of partnerships, the UN Secretary-General stated that:

> The United Nations and its member states are looking increasingly to the power of partnerships to achieve the millennium development goals and address other global challenges.... Such partnerships and alliances are the wave of the future in helping governments to pursue their development priorities and in eliciting the engagement of all those in a position to help make a difference (Davidi, 2013).

There may be a growing prominence of partnerships and voluntary initiatives within the UN context, but there is no escaping the definitive finding of the 2013 report of the UN's Joint Inspection Unit which reviewed the management of implementing partners within the UN:

Partnerships with public and non-public entities have become essential for most United Nations system organizations in pursuing their mandates. In many such partnerships, the organizations assign the implementation of programme activities to implementing partners (IPs).... Over all, the volume of United Nations resources entrusted to IPs is significant. Some organizations expend over half their annual budgets via IPs.... There is no clear definition for partnerships in general or implementing partners in particular. United Nations system organizations use various terms and definitions depending on their business models and type of intervention" (UN Joint Inspection Unit, 2013, Executive Summary, p. iii).

The Joint Inspection Unit's finding of a lack of a clear, system-wide, operational definition of partnerships within the UN, and the point that different UN entities use a diverse range of approaches and terms in the implementation of partnerships, provides the necessary cautionary backdrop for the discussion on locating energy access and climate change within the diverse range of partnerships and voluntary initiatives. The surging popularity of voluntary and collaborative initiatives that have been sanctioned as a result of UN globally agreed sustainable development outcomes, in the midst of a lack of definitional clarity on partnerships at the UN, has only made it that much harder to understand and differentiate amongst the wide variety of voluntary partnerships and actions.

The main focus of this chapter is to examine the wide range of voluntary initiatives that have emerged as a consequence of UN global outcomes related to sustainable development in the past two decades. The aim is not to provide a prescriptive or analytical tool to determine the efficacy of individual or groups of voluntary initiatives, but instead to provide a better understanding of the panoply of diverse voluntary efforts undertaken within the context of the UN. This chapter will provide evidence that voluntary initiatives on sustainable development under the aegis of the UN are varied in terms of scope, coverage, actors involved, and implementation modalities, and could benefit from a more detailed, comprehensive, and transparent framework that could allow clearer differentiation and accountability amongst and across diverse initiatives. Again, the objective is not to analyze all or any particular sustainable energy- and climate change-related voluntary initiatives, but to explain the context in which these initiatives have emerged and are located. The chapter will not therefore discuss all multi-stakeholder and bilateral partnerships on energy and climate change, nor will it provide case studies of individual PSDs or voluntary commitments (VCs), but instead it will reference the role of UN-convened or registered PSDs and VCs on energy access and climate change.

Partnerships for sustainable development have been part of the intergovernmental development lexicon for close to two decades, starting with 1992 Agenda 21, and more recently with the 2012 FWW outcome document. However, within the broad catch-all of voluntary efforts focused on sustainable development that have emanated as a consequence of intergovernmental negotiations on sustainable development, a new variant – VCs – emerged during the 2012 Rio+20 Summit. Based on a close examination of UN-led globally negotiated sustainable development outcomes and processes ranging from the 1992 Agenda 21, the Commission on Sustainable Development (CSD), the 2002 JPOI and the 2012 FWW, this chapter attempts to differentiate amongst a wide variety of voluntary efforts and partnerships that have emerged within the context of the UN. It discusses the emergence of PSDs and differentiates between them and the newer concept of VCs as well as the Global Partnership for Sustainable Development (GPSD) and the Global Partnership for Development (GPD). It provides an overview of key energy access and climate change-related PSDs and VCs that have been referenced within the context of the UN, as well as those that have been convened outside the aegis of the UN. The overall aim is to provide an improved understanding of the different categories of voluntary efforts and the need for a clear, transparent, and comprehensive framework that would allow for keeping track of diverse voluntary efforts, including any on energy access for poor and climate change.

4.2   The variegated world of PSDs, GPSD and VCs within the UN context: Making the case for conceptual and definitional clarity

A globally acceptable definition of the term "partnerships" can be found in the 2005 UN General Assembly resolution that was expressly entitled "Towards global partnerships". This UNGA resolution stressed that

> …partnerships are voluntary and collaborative relationships between various parties, both public and non-public, in which all participants agree to work together to achieve a common purpose or undertake a specific task and, as mutually agreed, to share risks and responsibilities, resources and benefits.

What is also important to highlight is that this resolution was explicit in recognizing "the importance of the contribution of voluntary partnerships to the achievement of the internationally agreed development goals, including the Millennium Development Goals" while re-emphasizing that "they are a complement to, but not intended to substitute for, the commitments made by Governments with a view to achieving these goals" (UNGA, 2006, p. 3). In other words, the voluntary and complementary nature of partnerships puts them in a different category – complementary and non-substitutable for governmental action related to globally agreed goals – but the trickier part is how and whether voluntary efforts emanating as a result of key UN global outcomes on sustainable development are measured and accounted for, especially when governments vary in terms of being donors and/or stakeholders in these voluntary efforts.

This complicated linkage alluded to in the UNGA resolution between non-legally binding, voluntary partnership efforts contributing to globally agreed outcomes and legally binding goals needs to be carefully considered, particularly in terms of the UN-led quest for a shared post-2015 development agenda. Because, if partnerships are defined as voluntary and complementary to globally agreed goals/outcomes, but also critical to the overall implementation of these agreed goals/outcomes, then what are the processes by which such partnerships can be evaluated in terms of their concrete contributions to agreed goals/outcomes. If, on the other hand, there is no globally acceptable accountability and evaluation framework by which voluntary initiatives can be assessed close to 20 years after the inception of voluntary partnerships, then how exactly have varied voluntary efforts been documented and verified as contributing to globally agreed goals/outcomes?

To begin with it is important to recognize that the world of UN affiliated multi-stakeholder voluntary initiatives is vibrant and variegated, but there is a need for clearer differentiation amongst diverse kinds of voluntary efforts. For instance, PSDs highlighted under the aegis of UN CSD-related partnership fairs and fora in the years since the adoption of 1992 Agenda 21 cannot be conflated with the concept of VCs that have been more recently referenced in the 2012 FWW outcome.

In general, there also appears to be considerable definitional and institutional ambiguity associated with differentiating amongst the more all-encompassing goal of a GPD, which was first given official prominence and enshrined as Millennium Development Goal 8 (MDG-8), and the GPSD, which was first mentioned in Agenda 21. The idea that global partnership action has been reflected in two broad umbrella categories needs to be carefully considered, because the concept of global partnership has recently re-emerged in the 2012 FWW outcome document and the ongoing UN discussions on SDGs, which are supposed to replace the MDGs in the post-2015 development agenda.

It is important to note the difference between the GPSD (which, as this chapter will show, has been referenced in both the Agenda 21 and the 2012 FWW but has not been acted upon in any systematic fashion at the global level) and the GPD, which, as MDG 8, has a pressing 2015 deadline for achievement and which has been briefly touched upon in Chapter 3. According to a recent UN report of the MDG Gap Task Force, whose intended role is to keep track of the progress and point out gaps related to the MDGs, entitled *The Global Partnership for Development: Making Rhetoric a Reality* (UN,

2014), there are several broad "targets" that were identified in association with the GPD-MDG-8 (UN, 2014, pp. ix–x):

> Target 8.A: Develop further an open, rule-based, predictable, non-discriminatory trading and financial system. Includes a commitment to good governance, development and poverty reduction – both nationally and internationally.
>
> Target 8.B: Address the special needs of the least developed countries. Includes tariff and quota free access for the least developed countries' exports; enhanced programme of debt relief for heavily indebted poor countries (HIPC) and cancellation of official bilateral debt; and more generous ODA for countries committed to poverty reduction.
>
> Target 8.C: Address the special needs of landlocked developing countries and small island developing States (through the Programme of Action for the Sustainable Development of Small Island Developing States and the outcome of the twenty-second special session of the General Assembly).
>
> Target 8.D: Deal comprehensively with the debt problems of developing countries through national and international measures in order to make debt sustainable in the long term.
>
> Target 8.E: In cooperation with pharmaceutical companies, provide access to affordable essential drugs in developing countries.
>
> Target 8.F: In cooperation with the private sector, make available the benefits of new technologies, especially information and communications.

In terms of the GPD, there are series of specific indicators that function as the measures by which the UN can keep track of the progress made towards the achievement of MDG-8 targets, and these indicators primarily relate to official development assistance (ODA) and debt relief such as net, country-specific, bilateral, aggregated and disaggregated, tied and untied ODA figures as well as figures for debt relief and debt service ratios, but there is no link between the GPD targets and indicators and environmental sustainability, which is separately enshrined as MDG-7. At a special 2010 High Level Plenary Summit of the UNGA, the UN Millennium Project found that donor countries pledged to increase ODA to $146 billion in 2010, but that 2010 ODA levels are projected to be about $126 billion and that "this shortfall in aid affects Africa in particular", and that "only 5 donor countries have reached or exceeded the UN target of 0.7 per cent of gross national income Denmark, Luxembourg, the Netherlands, Norway and Sweden", with the largest donors by volume "in 2009 being the United States, France, Germany, the United Kingdom and Japan". It also found that, although debt burdens on developing countries have eased, "existing major debt relief initiatives are coming to an end, and a number of low-income and small middle-income countries are in or at risk of debt distress" (UN Millennium Project, 2010, p. 1). The issue is that that some of the "targets" read more like aspirational or indicative ideas has proved hard to translate into concrete and effective implementation and programmatic action. The most recent UN report on the status of progress towards MDG-8, however, is much more cautionary, noting that there was "difficulty identifying areas of significant new progress and for the first time there are signs of backsliding" and that "fewer MDGs will be reached in fewer countries as a result", and pointing out "that the commitment to strengthen international cooperation to address sustainable development for all made at 2012 Rio+20 would only be credible if promises made to achieve the 2015 MDGs was achieved" (UN, 2014, Executive Summary, p. xi).

With the clock ticking towards the 2015 MDG deadline, the UN Special Event to Follow-Up Efforts to achieve the MDGs, held in September 2013, stressed the need to strengthen the existing global partnership for development encapsulated as MDG-8 and work towards a post-2015 development agenda that has a single framework and set of goals that are universal and applicable to all countries (UNGA, 2013, pp. 3–4). But here the question is how exactly do the plethora of voluntary PSDs interact or fit in with the broader globally agreed goal of implementing a GPD, especially as work on advancing the GPD has not progressed adequately, and, more importantly, how will any future work to advance the GPD be reflected or differentiated in any future GPSD? While other MDGs have secured global attention and action, actual progress on MDG-8 – the GPD that could provide foci for poverty reduction partnerships including energy access for the poor – has been disappointing. In addition, it is hard to distinguish between actual and potential partnerships focused on addressing MDG-7 versus partnerships related to the overarching GPD/MDG-8, because of an overall lack of definitional and accounting clarity on partnerships.

To make matters more challenging when it comes to the issue of global partnerships, in addition to the GPSD and GPD, there is another new addition – the Global Partnership For Effective Development Co-operation (GPEDC) – launched in Busan, South Korea, as a follow-up to a longer trajectory of global efforts to make development cooperation more effective, ranging from the Monterrey Consensus of 2002 to the Paris Declaration on Aid Effectiveness (2005) and the Accra Agenda for Action in 2008. The stated aim of this GPEDC to "help drive progress and support the implementation of the global development agenda that will follow the Millennium Development Goals target year of 2015" and, according to the GPEDC, "160 governments and 46 organizations have endorsed the Busan agreement which includes 4 key principles – ownership of development priorities by developing countries, focus on results, inclusive development partnerships, transparency and accountability to each other" (see www.effectivecooperation.org/).

There are overlaps in terms of the stated aim and mission of the GPEDC, and the broader more encompassing roles envisaged for the GPSD and the

GPD, both of which are referenced in globally agreed UN outcomes. Unlike the GPSD and the GPD, the GPEDC emanated as an Organisation for Economic Co-operation and Development (OECD) initiative, and currently the United Nations Development Programme (UNDP) and the OECD work jointly to provide support for the functioning of the GPEDC. With the first official meeting of the GPEDC completed in April 2014, the very real concern is whether two overlapping and competing global partnerships could be creating a new set of implementation and delivery challenges, given a limited and finite set of resources (financial and capacity) focused on development. Glennie (2014) points out that the existence of two global partnerships could "negate attempts to monitor and improve development co-operation" and has stated that the "big question is, how can the Global Partnership for Effective Development Co-operation, with its more limited focus on aid and finance, fit with the more ambitious and overarching "global partnership for effective sustainable development". The actual challenges associated with delivering on the 2015 MDG deadline are made explicit in a 2014 document, entitled *Making Development Cooperation More Effective*, which finds that the "glass is half full" and that "much more is needed to implement commitments by 2015" (OECD/UNDP, 2014, p. 23).

In addition to possible overlaps in distinguishing between three different variants of global partnerships, there is the much bigger challenge of how voluntary efforts, including both VCs and PSDs, are being accounted for. A more detailed discussion on PSDs and VCs registered within the Sustainable Development in Action (SD in Action) Registry is provided later in the chapter, which highlights the need for accuracy in reporting of voluntary initiatives as a key concern.

Given the plethora of voluntary initiatives and their increasingly important role in delivering on globally agreed outcomes, it becomes useful not just to differentiate between PSDs and VCs, but to differentiate between both of these categories of voluntary-based efforts and the role of the globally agreed GPD and/or future role of the GPSD. Any future global partnership, whether it is named the GPD or the GPSD, will, by default, need to be driven by voluntary partnerships, and could allow for VCs to fulfill its broad objectives, but at this point in time and based on a review of agreed global outcomes relating to sustainable development, it is not clear how and whether such implementation linkages have even been made. It is also important to recognize at the outset that the implementation trajectories and implementation records of PSDs has, for the large part, been difficult to comprehensively access and evaluate, primarily due to the fact that the evolution of these voluntary PSDs has tended to be somewhat ad hoc, with results and records being largely self-reported and not subject to any detailed and verifiable monitoring and reporting framework.

In terms of delivery of concrete outcomes and actual services, PSDs aimed, for instance, at delivering energy access should therefore also be considered somewhat distinct from coalition partnerships, which are discussed by Joyner

(2005) as like-minded coalitions of groups, in which partnership coalitions can be used for the regime construction, management, and regulation of global environmental issues. An example of the formalization of NGO coalitions created as a consequence of the follow-up to the 1992 UN Conference on Environment and Development (UNCED) is the Committee of the Conference of NGOs in Consultative Relationship with the United Nations (CONGO). The purpose of CONGO is to monitor the implementation of commitments included in the Agenda 21 and other agreements adopted at UN meetings that pertain to sustainable development, but the CONGO does not constitute a voluntary PSD, nor was it intended as one. While policy networks and networks for action can be devised as partnership activities, from the immediate perspective of this chapter, only those voluntary partnerships related to energy and climate change that have been registered with the UN will be touched upon. Examples of other kinds of non-UN-led voluntary alliance and coalitions for dialogue and/or action on climate change and energy, such as the Major Economies Forum on Energy and Climate Change (MEF) and the Climate and Clean Air Coalition, are discussed separately in Chapter 5, which looks at the role of non-UN-led efforts and initiatives that attempt to address the nexus between energy and climate change issues.

Partnerships for sustainable development cannot be neatly conflated with what is more commonly referred to as public–private partnerships (PPPs), which are very much part of the lexicon of development efforts at all levels (local, national, and global) and include participation from a range of civil society actors, but most importantly involve a contractual agreement between state and non-state actors. In other words, PSDs, while ostensibly involving both state and non-state or civil society actors, can also involve an amalgam of various non-state actors and completely exclude any state actors. Broadly speaking, PSDs within the UN context are therefore voluntary, multi-stakeholder, collaborative efforts relating exclusively to sustainable development concerns and emanating under the broad aegis of the UN, though not subject to any UN-related contractual arrangements. PPPs can involve, by default, a variety of state (at different levels) and non-state actors, but are not envisaged as excluding some form of state involvement and therefore also include some kind of defined contractual arrangements between state and non-state actors involved in the particular PPP.

For instance, PPPs have in the past decade become an important route/ vehicle for governments to seek extra-budgetary resources in key areas of infrastructure development and the provision of needed services. PPPs in this regard are much more scripted and governed by a clear set of rules and procedures, including rules for consultation, preparation, procurement, and management. The role of PPPs has been both highly promoted and also contested in individual countries, as demonstrated, for instance, by the Indian government, which has a clearly stated interest in engaging with PPPs, as represented by its ongoing search for finalizing the national PPP rules and policies. In the case of India, a national PPP capacity-building program was launched in the Finance Ministry in 2010 and is anticipated to train 10,000 senior and middle-level government officials over the next 3 years (see the Public Private Partnerships in India website: www.pppinindia.com).

Although a lot of attention has been focused on PPPs, they have been hard to define concisely and evaluate precisely because of the sheer diversity of coverage, scope, institutional arrangements, and actors involved. Linder and Rosenau (2000), for example, have defined PPPs more broadly as "the formation of cooperative relationships between government, profit-making firms, and non-profit private organizations to fulfill a policy function" (p. 5). Colverson and Perera's (2012) landmark report on PPPs in the area of hybrid financing for sustainable development provides perhaps the most concise summation of the problem in which they point out that PPPs are a "generic term being used" for a variety of different types of contractual agreements "between the State and the private sector for the purpose of public infrastructure development and services provision"(p. 2). They point out that there is "no one single, concise definition of PPP" and that defining PPPs accurately "is problematic because by nature it is a contextual concept, responding to the institutional, legal, investment and public procurement settings of different jurisdictions, whilst also considering the contextual nature of individual agreement" (Colverson and Perera, 2012, p. 2).

There is also ample room for confusion in terms of defining the concept of partnerships, because a review of the existing literature on partnerships related to sustainable development reveals a considerable amount of definitional ambiguity in terms of distinguishing between partnerships focused on sustainable development emanating as result of an intergovernmental outcome such as the 2002 JPOI, and the broader conceptual role of PSDs in terms of their impacts on environmental governance. A review of the current academic literature reveals that more focused attention is needed to differentiate amongst the array of well-intentioned, yet not so well understood and not so easily verifiable, voluntary efforts at sustainable development, including both PSDs and VCs that have been undertaken within and outside the UN context. Analyses on the role of partnerships in the broad area of environment and/or sustainable development has focused on understanding whether these partnerships can be seen as a new and influential form of environmental governance (Glasbergen et al., 2008) or whether these "public-private partnerships for the Earth" constitute a hybrid governance arrangement (Adonova, 2010). Glasbergen et al. (2008) provide three different but not necessarily compatible perspectives on the role of, and processes associated with, PSDs, which include the processes by which partnerships work as collaborative efforts; the impacts of partnerships and how they can influence change;

and the broader governance implications of partnerships on decision-making. But here the distinctions between partnership actions related to sustainable development, the GPSD and the GPD encapsulated by MDG-8, as well as the more recent emergence of voluntary commitments related to particular sustainable development related objectives remain to be clarified.

The most comprehensive analysis of PSDs has been provided by Pattberg *et al.* (2012), who note that one of the noteworthy outcomes of the 2002 World Summit on Sustainable Development (WSSD) was its promotion of PSDs, which can be viewed as a distinct and novel contribution to global governance. Combining a case study approach of key partnerships as well as an assessment of the Global Sustainability Partnerships Database, they not only provide an in-depth review of the effectiveness of the post-WSSD decade of partnerships, but also compare across thematic/sectoral and geographic partnerships, including the role of energy- and water-related partnerships, as well as those focused in China, India, and Africa. But here the differences between voluntary partnerships and voluntary commitments and the nature and scope of any linkages and/or the related implications of both voluntary partnerships and commitments are not discussed in any great detail.

From the early inception of the concept of partnerships, including both PSDs and the GPSD, there was no stated attempt to provide or devise a framework, database, or registry that could keep account of these varied partnership actions in terms of the agreement contained in either Agenda 21 or the JPOI. Agenda 21 contained 53 different references to databases and one to a land registry, but none of these relate to the issue of partnerships or to the implementation of the GPSD. Meanwhile, the JPOI contained no mention of any kind of registry, and included only three references to the need for databases, including water resources-related databases, integrated databases on development hazards in relation to heath, and the accurate databases for developing countries in relation to environmental protection. Despite a 20-year trajectory of voluntary efforts, including both PSDs and PPPs, focused on range of sustainable development issues and an ostensible GPSD, the first mention of a registry is contained in the 2012 FWW. The FWW contains not one single mention of any kind database but includes three references to a registry on VC in its concluding paragraph (para. 283), which is discussed at length later in this section. There is no reference to a partnerships registry in the FWW.

The key concern that needs to be carefully considered is the promotion of diverse voluntary efforts that are largely self-reported and only relatively recently included within the framework of two separate UN registries. There is currently no comprehensive reporting or accountability framework that can allow for comparisons and linkages between and amongst a diverse range of voluntary initiatives on sustainable development. The lack of clear accountability framework for PSDs, VCs, and the overall implementation of the GPSD does pose verification challenges for any future shared UN-led post-2015 development agenda that is fueled by voluntary partnerships and efforts.

Based on the growing role of such voluntary efforts aimed at sustainable development, in particular those focused on sustainable energy and climate change concerns, and their ostensible role in fulfilling the implementation of agreed global outcomes relating to sustainable development, it is clear that the different kinds of voluntary efforts need to be further examined and better understood. As a first step towards understanding the various mechanisms for voluntary efforts that have been broadly shepherded by the UN, it is useful to begin by differentiating between different categories of efforts that have emerged as a consequence of intergovernmental negotiations and agreements relating to sustainable development over the past two decades, namely PSDs, the GPSD, the GPD, and VCs.

The emergence of each of these different categories of voluntary efforts based on a review of key UN-related intergovernmental outcome documents on sustainable development, including the 1992 Agenda 21, the CSD process and the 2012 FWW, is provided in the following section. Based on a careful review of key UN-led globally agreed outcomes, Section 4.3 looks at how and whether the PSDs and the GPSD are referenced in relation to each other, and also examines how the newest variant in the category of voluntary efforts – VCs – mentioned for the first time in the FWW, can be understood.

4.3   Tracing the emergence of PSDs, GPSD, and VCs within key global sustainable development outcomes from 1992 to 2014: Locating energy access and climate change

The main outcome document of the WSSD – the 2002 JPOI – emphasized the role of partnerships as a mechanism for the implementation of sustainable development which was distinct from intergovernmental action. In its opening paragraphs, the JPOI was clear that "the implementation of the outcomes of the Summit" should "involve all relevant actors through partnerships, especially between Governments of the North and South, on the one hand, and between Governments and major groups, on the other, to achieve the widely shared goals of sustainable development", and it also reiterated the Monterrey Consensus' claim that "partnerships are key to pursuing sustainable development in a globalizing world" (UN, 2002, para. 3, p. 2). Interestingly, Spalding-Fecher *et al.* (2005) have argued that the absence of leadership from key countries prevented substantive agreement on both renewable energy targets and a program to promote energy access, and that the WSSD achievements were more focused on enabling activities rather than concrete institutional action on sustainable energy and energy access.

Partnerships for sustainable development have been officially heralded by the UN as one of the most significant outcomes – type 2 outcomes due to their non-legally binding and voluntary nature – of the 2002 WSSD, where, according to the UN, 196 such partnerships were launched in the lead-up to and after the Summit (see the UN Sustainable Development Knowledge Platform website, https://sustainabledevelopment.un.org/sdinaction/wssd). In reality, though, the concept of PSDs emerged a decade before the 2002 WSSD discussed a wide variety of partnership actions in outcome document – the JPOI. The concept of partnerships within the context of UN-led intergovernmental negotiations on sustainable development can be traced directly back to the 1992 UNCED, where PSDs and the GPSD were both seen as a means of assisting with the implementation of Agenda 21, and also in Programme for the Further Implementation of Agenda 21, agreed to in 1997. Within the UN context, the practical locus for partnership action related to a range of sustainable development issues has been the UN's CSD. As discussed in Chapter 3, the UN's CSD was established in 1992 as a follow-up institutional outcome to the 1992 Rio Summit and its Agenda 21 outcome document. The CSD has recently been replaced by the convening of high-level political forum on sustainable development (HLPF), which was the institutional response of the 2012 Rio+20 Summit and which met for the first time in June 2013 and has been charged with delivering on the outcome of the 2012 FWW.

It was, in fact, Agenda 21 in its opening preambular paragraph that first called attention to the need for a GPSD by stating emphatically that no nation could achieve sustainable development on its own but that "together we can – in a global partnership for sustainable development" (UN, 1992, para. 1.1). It also noted that that implementation of its various program areas marked "the beginning of a new global partnership for sustainable development" (UN, 1992, para. 1.6). These two references constitute the first mention of the GPSD in a globally agreed outcome document related to achieving the broad goals of sustainable development. Agenda 21 discussed the role of a GPSD as a new global partnership, and also mentioned the role of partnerships that are more thematic or sectoral in focus as being important to the future implementation of its agreed outcomes.

There are 45 references to the concept of partnership or partnerships in Agenda 21's chapters, dealing with a range of examples such as the need for a new global partnership for international cooperation to accelerate sustainable development in developing countries (para, 2.1); the encouragement of partnerships in managing land resources for human development (para. 7.30 (d)) and the creation of an enabling policy environment supportive of the partnership between the public, private, and community sectors in human development (para. 7.77 (b)); the promotion of popular participation rooted in the concept of partnership related to desertification control and drought management (para. 12.55); maintenance and establishment of partnerships with NGOs and other groups working in watershed development (para. 13.8 (c)); and the new opportunities for global partnerships in biotechnology (para. 16.1). The role of partnerships is perhaps most notably mentioned in the section of Agenda 21 dealing with the role of major groups, including women, children and youth, indigenous people, NGOs, local authorities, business and industry, farmers, and scientific and technological community.

The dual aspect of implementing partnership actions in terms of having both an overarching global partnership and individual, sectoral, or thematically focused partnerships contained in Agenda 21 is important to highlight and contrast with the globally agreed outcomes documents emanating from 2002 WSSD and the 2012 Rio+20 Summit. Unlike Agenda 21, the JPOI does contains not any reference to the concept of a GPSD, nor does it contain any reference to the GPD, which is surprising considering that the entire focus of the JPOI was on sustainable development, and also given that MDG-8, focusing on the GPD, was being heavily promoted during the time of the WSSD.

The CSD was the intergovernmental negotiating forum that resulted as a direct outcome of Agenda 21 and it provided the official intergovernmental guidance and inputs on the scope of implementation and direction of partnership action relating to sustainable development in the post-Agenda 21 era leading up to the 2002 JPOI and again in the lead-up to the 2012 FWW. It was the early sessions of the CSD that set the momentum and tone of voluntary partnership action. For instance, the issue of PSDs was highlighted at the first session of the CSD, which was held a year after the 1992 Rio Summit. The chairman's summary of the 1993 high-level meeting of CSD-1 noted that ministers had "reconfirmed their commitment to the growing global partnership for sustainable development among nations" and "highlighted the need for further development of such partnerships as that upon which Agenda 21 had been built" (UN Economic and Social Council [UNECOSOC], 1993, p. 24). Although there is no other mention of the role of partnerships in this first session, the earliest and most detailed references to the role of partnerships as voluntary and non-binding agreements between governments and civil society stakeholders, including the private sector, emerged in the second session of the CSD. The final report of CSD-2 contained six references to the role of partnerships, and perhaps the most significant reference contained in CSD-2 was the recommendation that "States and international organizations consider the use of partnerships with business and non-governmental communities leading to non-legally binding agreements as a first step in the preparation of international regulations" (UNECOSOC, 1994, p. 6).

CSD-2 also encouraged all "major groups", which is standard UN parlance for a range of civil society actors, to engage in creating multi-stakeholder partnerships and to carry out concrete partnership projects (UNECOSOC, 1994, p. 7). So early on there was this idea that voluntary partnerships could lead to non-legally binding outcomes and that these non-binding agreements/outcome could also simultaneously serve as the start of the process of globally acceptable regulations but there was no institutional analog or framework that could account or track progress towards these outcomes.

The third session of the CSD (CSD-3) specifically "encouraged partnerships among Governments, intergovernmental organizations and the major groups in jointly organizing future inter-sessional activities and other relevant meetings" and can be seen as the signal for future convening of partnership activities within the context of official CSD meetings (UNECOSOC, 1995, p. 5). What is interesting to note is that these early sessions of the CSD mentioned the role of partnerships but also agreed on a broader concept of "voluntary initiatives and agreements". For instance, CSD-6 pointed to the "potential value of a review of voluntary initiatives and agreements", and for the first time invited UNDESA to work with representatives of industry, trade unions and non-governmental organizations and the UN Environment Programme (UNEP) and the UN Industrial Development Organization (UNIDO) in preparing a review of voluntary initiatives that would examine "how voluntary initiatives and agreements could contribute to the future work of the Commission" (E/CN.17/1998/20, para. 17). Meanwhile the following session, CSD-7 stressed:

> ...the need for better understanding of the possible impacts of voluntary initiatives and agreements on developing countries including through periodic reports through the CSD Secretariat on steps they have taken or progress they have made in assisting developing countries in understanding and making use of, as appropriate, the lessons to be learned from the use of voluntary initiatives and agreements (E/1999/29-E/CN.17/1999/20, para. 1 (f)).

The initiation of the concept of partnerships fairs can be traced back to CSD-11, which called for the active involvement in partnership-related and capacity-building activities at all levels, including "the partnerships fairs and learning centers organized as part of the meetings of the Commission" (E/CN.17/2003/6, para. 20 (e)). According to the UN's Special Report on sustainable development in action, the concept of "partnership", in fact, placed emphasis on associations of stakeholders, which met specific criteria established by CSD-11. In February 2004, UNDESA launched the CSD's Partnerships for Sustainable Development website, which included a partnerships database from the 12th session (2004) onwards, and the partnership fair became an essential element of the CSD negotiations process.

As discussed in Chapter 3, energy for sustainable development-related issues were the main focus of CSD-14 and CSD-15, with aim that the 2007 CSD-15 would provide policy recommendations on a variety of energy-related areas, including energy access and the linkages between household air pollution, poverty and ill-health. The political tensions and inability of governments to reach a consensus agreement in terms of agreeing on final summary text of CSD-15 has been highlighted in Chapter 3, but what is important to note here is that the lack of consensus and agreement at CSD-15 has serious implications going into the future in terms of energy-related guidance and inputs on partnerships and initiatives, including the Sustainable Energy for All (SE4All) initiatives sponsored by the UN Secretary-General. Rehfuess (2007) pointed out that CSD-15 should be seen as a "missed opportunity" for concrete policy actions and recommendations focused on addressing the serious development challenge of how to increase access to clean, efficient, and safe energy services for the poor. The World Health Organization (WHO) has argued that since the outcome of CSD-15 was not an agreed policy document, the Chair's summary, which contained several clear recommendations on indoor air pollution and health, including, for example, the need to replicate and scale up "successful approaches and best practices and partnerships to reduce indoor air pollution, such as the Partnership for Clean Indoor Air, with a priority focus on practical initiatives that improve air quality", remains unresolved (WHO, 2007).

The concept of PSDs was given special prominence in the lead-up to the 2002 WSSD. Although voluntary multi-stakeholder partnerships were emphasized at the 2002 WSSD, the guiding principles for PSDs, also referred to as type 2 outcomes (in contrast to type 1 outcomes which are globally agreed outcomes and goals), were contained in an Annex document prepared by the CSD in the lead-up to the 2002 WSSD. In this document, PSDs are defined as:

> ...specific commitments by various partners intended to contribute to and reinforce the implementation of the outcomes of the intergovernmental negotiations of the WSSD (Programme of Action and Political Declaration) and to help achieve the further implementation of Agenda 21 and the Millennium Development Goals (UN CSD, 2002, Annex, p. 1).

The PSDs "are voluntary, [of a] 'self-organizing' nature" and "based on mutual respect and shared responsibility of the partners involved, taking into account the Rio Declaration Principles and the values expressed in the Millennium Declaration" (UN CSD, 2002, Annex, p. 1). So here again were additional references in which PSDs were seen as reinforcing and assisting in the implementation of Agenda 21, the MDGs and the JPOI.

Often referred to as type II voluntary initiatives, (to distinguish them from the type I or the politically negotiated agreements emanating from the WSSD), these voluntary partnerships are conceived as facilitating new and complementary forms of cooperation among diverse stakeholders. As discussed in Chapter 3, many questions remain in relation to these voluntary type II partnerships which are not answered in the JPOI or the FWW, such as exactly how type I outcomes (those contained in the JPOI and the FWW) and type II partnerships can be linked and what procedures best facilitate the broader implementation and scaling up of PSDs.

Although the JPOI does not mention the GPSD, it does contain 51 references to a wide variety of partnerships which are found throughout the outcome document. Rather than specifically invoking the concept of PSDs in the JPOI, considerable emphasis is placed on the role of PPPs aimed at a diverse range of sectoral/thematic issues related to sustainable development, which could explain the definitional ambiguities and conceptual challenges associated with differentiating between PSDs as they emerged in Agenda 21, and more recently in the FWW, and PPPs aimed at sustainable development in the JPOI.

The JPOI, for instance, highlights the role of PPPs related to various areas such as:

- PPPs "aimed at increasing agriculture production and food security" (para. 7 (j), p. 3);
- PPPs and other forms of partnership that give priority to the needs of the poor relating to integrated water resource management issues (para. 26 (g), p. 15);
- active promotion of "corporate responsibility and accountability" through PPPs and other measures (para. 49, p. 31);
- "public–private multisector partnerships" for "safe water, sanitation and waste management for rural and urban areas in developing countries" (para. 53 (l), p. 32).

Additionally, the JPOI also referenced a whole host of other kinds of partnership-based activities, including "community-based partnerships linking urban and rural people and enterprises" (para. 7 (k), p. 3); the role of "innovative financing and partnership mechanisms" for the implementation of water and sanitation goals, (para. 8 (f), p. 5); workplace-based partnerships and programs, including training and education programs (para. 18 (d), p. 8); "partnerships between international financial institutions, bilateral agencies and other relevant stakeholders to enable developing countries" in the sustainable use and management of fisheries (para. 31 (g), p. 18), "partnerships with both private and public sectors, at all levels in the area of sustainable tourism" (para. 43 (a), p. 26); and strengthened "partnerships for financial resources and capacity building in the areas of forests" (para. 45 (f), p. 29).

In terms of energy access for the poor, perhaps the clearest invocation of the role of PPPs in the JPOI can be found in paragraph 9, which calls for assistance and facilitation by developed countries, "including through public–private partnerships, the access of the poor to reliable, affordable, economically viable, socially acceptable and environmentally sound energy services, taking into account the instrumental role of developing national policies on energy for sustainable development" (para. 9 (g), pp. 5–6). As discussed in chapter 3, paragraph 20 essentially serves as the energy focal area of the JPOI, with its 23 different energy-related recommendations. In the opening segment of this key JPOI paragraph, countries are urged to implement actions within the framework of CSD-9, "including through public-private partnerships, taking into account the different circumstances of countries... in the field of access to energy, including renewable energy and energy efficiency and advanced energy technologies, including advanced and cleaner fossil fuel technologies" (para. 20 (t), p. 12).

In contrast to the call for PPPs in a variety of energy for sustainable development areas, the JPOI does not contain any references to the role of partnerships in addressing any specific climate change mitigation objectives. It includes only one minor mention of the need for partnerships that could enable SIDS to address their adaptation needs (para. 58 (j), p. 35). However, it is important to highlight an interesting linkage among PPPs, energy access and climate change in the JPOI, namely its call to: "Reduce respiratory diseases and other health impacts resulting from air pollution with particular attention on women and children" via various mechanisms which included the strengthening of regional and national programs through PPPs in developing countries, the phasing out of lead in gasoline, strengthened support of efforts for the reduction of emissions through the use of cleaner fuels and modern pollution control techniques; and assisting in the provision of affordable energy to rural communities in developing countries, "particularly to reduce dependence on traditional fuel sources for cooking and heating, which affect the health of women and children" (para. 56 (a–d), p. 33).

The overall importance accorded to partnerships was made explicit in the JPOI, which listed partnership initiatives for the implementation of the WSSD outcomes as one of three key actions (along with streamlining the international sustainable development meeting calendar and making full use of information and communication technologies) to be undertaken to promote the effective implementation of Agenda 21 at the international level (para. 156, p. 60).

Unlike the JPOI, which did not include any reference to a GPSD or to any kind of global partnership, but promoted the general concept of partnerships as a vital means of implementation of the WSSD, the FWW does contain one reference that revived and recalled Agenda 21's more frequent invocation of the GPSD. More than 20 years after the adoption of Agenda 21 and in the face of the looming 2015 deadline for the achievement of the MDGs, the 2012 FWW contains only two references to the concept of a "global partnership". The first reference has nothing whatsoever to do with either a GPD or a GPSD, and instead is focused on the Istanbul Programme of Action for least developed countries' (LDCs) priorities for sustainable development and "defines a framework for a renewed and strengthened global partnership to implement them" (UNGA, 2012, para. 34, p. 7). Interestingly, the FWW does contain a singular commitment to "re-invigorating the global partnership for sustainable development" that was "launched in Rio in 1992", and here the FWW points to the need to commit to work with major groups and other stakeholders in addressing implementation gaps related to sustainable development (UNGA, 2012, para. 55, p. 10).

Curiously, the FWW has fewer overall references to the concept of partnership in general – 25 references – in comparison to the JPOI. There are, however, new categories of partnership activities sanctioned within the FWW, such as the New Partnership for Africa's Development (NEPAD), which was a specific and unique outcome of the post 2002 WSSD time frame. The FWW included, for instance, three references to the NEPAD, two references to a Collaborative Partnership on Forests and a singular reference to a voluntary regional Green Bridge Partnership, none of which are contained in the JPOI, in addition to the promotion of partnerships in thematic issues related to sustainable development.

The FWW agreed to a more prominent role for the private sector and noted that "the active participation of the private sector can contribute to the achievement of sustainable development, including through the important tool of public-private partnerships". Accordingly, it supported "national regulatory and policy frameworks that enable business and industry to advance sustainable development initiatives, taking into account the importance of corporate social responsibility", and called upon the private sector to engage in responsible business practices, such as those promoted by the United Nations Global Compact (para. 46, p. 9). Now this is important to point out because the FWW did not reference any generalized partnership database or framework for accountability or implementation other than what is cited earlier in relation to private sector actions. So, for instance, even when the FWW acknowledged the role of "all stakeholders and their partnerships" in assisting "countries to learn from one another in identifying appropriate sustainable development policies, including green economy policies" (para, 64, p. 12), and encouraged "existing and new partnerships, including public-private partnerships, to mobilize public financing complemented by the private sector, taking into account the interests of local and indigenous communities when appropriate (para. 71, p. 13), there was no stated or explicit role for specific partnership databases or framework that could promote such learning or speed up mobilization of resources.

The FWW recognizes a wide array of partnership activities as playing an important role in promoting sustainable development, such as:

- partnerships among cities and communities and the need to "strengthen partnership arrangements and other implementation tools to advance the coordinated implementation of the Habitat Agenda" (para. 137, p. 27);

- "opportunities for decent work including through partnerships with small and medium-sized enterprises and cooperatives" (para. 154, p. 29);
- role of "international cooperation and partnerships" in relation to protection of biodiversity (para. 202, p. 39);
- "the importance of partnerships and initiatives for the safeguarding of land resources" (para. 207, p. 40);
- "strengthening capacities, including through partnerships, technical assistance and improved governance structures" for the sound management of chemicals and waste in LDCs (para. 215, p. 42);
- role of "existing public-private partnerships, and call for continued, new and innovative public-private partnerships" aiming to "enhance capacity and technology for environmentally sound chemicals and waste management, including for waste prevention" (para. 217, p. 42);
- need for "international educational exchanges and partnerships, including the creation of fellowships and scholarships to help to achieve global education goals" (para. 232, p. 45);
- role of "new partnerships and innovative sources of financing" in "complementing sources of financing for sustainable development" (para. 253, p. 48) and,
- support for "developing countries in particular the least developing countries in capacity building" related "resource-efficient and inclusive economies", by all UN agencies and international organizations through a variety of means including "promoting public-private partnerships" (para. 280, p. 53).

In contrast to both Agenda 21 and the JPOI, the FWW contained no mention of partnerships in relation to energy access for the poor and climate change mitigation and adaptation. In other words, the sole reference to partnerships for adaptation needs in SIDS that was in the JPOI is missing in the FWW. As discussed in detail in Chapter 3, unlike the 23 recommendations related to a broad range of energy for sustainable development issues, the FWW contained only five paragraphs referencing energy issues. But, in spite of the FWW explicit recognition that "access to sustainable modern energy services contributes to poverty eradication, saves lives, improves health and helps to provide for basic human needs" and its clear commitment "to facilitate support for access to these services by 1.4 billion people worldwide who are currently without them" given "that access to these services is critical for achieving sustainable development", the FWW did not mention partnerships of any sort – that is, PSDs or PPPs focused on the area of energy access (para. 125, p. 24). Instead of the concept of partnerships related to sustainable cooking and heating solutions, the FWW instead focused on the idea of "collaborative actions to share best practices and adopt policies as appropriate" and urged government to create "enabling environments that facilitate public and private sector investment in relevant and needed cleaner energy Technologies" (para. 127, p. 25). And finally, although the SE4All initiative had an extensive launch process in both global capital and at the UN, in the end, the FWW only took "note of the launching of the "Sustainable Energy for All" initiative by the Secretary-General" (para. 129, p. 25).

If PSDs were pushed to prominence in the 2002 WSSD, the 2012 Rio+20 Summit pushed the concept to VCs onto the global stage. As the discussion on Agenda 21 and CSD noted, the concept of voluntary initiatives was not new, but what the 2012 Rio+20 Summit did was provide a global forum and a showcase for stakeholders, ranging from governments to various civil society actors, to voluntarily pledge commitments in diverse areas related to sustainable development. But surprisingly, the actual FWW outcome does not contain a single reference to the term "voluntary commitment" or "voluntary commitments" and, in fact, the bulk of the 11 references to the word "voluntary" scattered in the FWW are in relation to the voluntary sharing of experiences, knowledge, and information. So there is no actual or concrete institutional guidance contained in the FWW that defines what exactly constitutes a voluntary commitment. The singular reference to this concept of voluntary commitments is contained in the concluding paragraph of the FWW, which focused on the "Registry of commitments" which welcomed "the commitments voluntarily entered into at the United Nations Conference on Sustainable Development and throughout 2012 by all stakeholders and their networks to implement concrete policies, plans, programmes, projects and actions to promote sustainable development and poverty eradication". This last paragraph of the FWW also invited "the UN Secretary-General to compile these commitments and facilitate access to other registries that have compiled commitments, in an Internet-based registry. The registry should make information about the commitments fully transparent and accessible to the public, and it should be periodically updated (para. 283, p. 53).

The need for a better resolution of the dichotomy between the complementary nature of varied, voluntary efforts and their stated crucial role in fulfilling the agreed outcomes of Agenda 21, the JPOI and the FWW becomes especially apparent in the absence of any globally accessible and comprehensive accountability framework by which implementation of varied voluntary efforts can be evaluated or examined. The 2012 FWW, in dealing with the institutional framework for sustainable development, called for the active engagement of the civil society and other stakeholders in the promotion of "transparency and broad public participation and partnership to implement sustainable development" (para. 76 (h), p. 15). But it made no specific mention of exactly how such transparency and accountability in partnership actions could be determined, nor did it lay out guidance for any future framework by which this could be undertaken.

The most globally acceptable definition of the concept of VCs is actually not contained in the FWW, but instead can be found in the special 2013 report prepared by UNDESA on the VCs and PSDs, which stated that: "The term 'voluntary commitment' came into use ahead of the Rio+20 Conference and emphasized the outcomes of associations rather than the associations themselves, although behind every voluntary commitment there is usually an association or Partnership." In further defining the concepts, the report stated that VCs and PSDs are "both terms" that "refer to voluntary, multi-stakeholder initiatives to promote sustainable development but the term "commitment" implies a shift in emphasis to implementation and outcomes rather than the social capital of associations" (UNDESA, 2013, p. 4).

Now what is curious about this definition is that it is the first time that a UN report alludes to fact that PSDs are not to be considered as outcomes related to the implementation of sustainable development objectives, including those contained in Agenda 21 and JPOI, but instead are to be seen as "the social capital of associations". This latter term is a new and unusual concept that is not derived from the agreed text on PSDs in Agenda 21, JPOI and the FWW, and this view of PSDs is contrary to the agreed language on PSDs to global sustainable development outcomes referenced earlier, in particular the JPOI. This view of PSDs as not being implementation-related also contradicts what is contained in the UNDESA website referencing partnerships at CSD-11/WSSD, which clearly stated: "At its 11th Session in May 2003, the Commission on Sustainable Development (CSD) reaffirmed that these partnerships contribute to the implementation of intergovernmental commitments, recognizing that partnerships are a complement to, not a substitute for, intergovernmental commitments" (UNDESA, Sustainable Development Knowledge Platform: https://sustainabledevelopment.un.org).

It would appear from this UN report that VCs are to be seen as more implementation-focused outcomes rather than PSDs but then the question that remains to be explained is why VCs are not clearly defined in terms of the FWW stating exactly what constitutes a VC, and why there is no explicit linkage between VCs being seen as complementary means for implementation of the FWW outcome, as was the case with partnerships in relation to Agenda 21 and the JPOI. This stated distinction between PSDs as being the social capital of associations and VCs as emphasizing implementation also raises concerns about how and whether these two different types of voluntary activities can be reconciled or linked in terms of overall implementation of sustainable development outcomes.

There is an important admission made in the UN special report on voluntary initiatives which needs to be carefully considered in terms of how information on partnerships and VCs are collected and the distinctions between these two categories of voluntary initiatives. In describing the 2004 CSD database on partnerships, the report stated that: "The web-accessible database contained information on registered partnerships based on voluntary self-reports from partnership focal points. These reports helped to keep the partnerships database up to date and formed the basis for summary reports prepared by the United Nations Secretariat" (UNDESA, 2013, p. 5). It is quite clear, therefore, that the information on voluntary partnerships was self-reported to UNDESA and these self-reports informed all analysis on PSDs linked to this database. The report raised a further concern that a total of 348 PSDs had been registered with the CSD as of February 1, 2010, but a "review of the partnerships ahead of the Rio+20 conference revealed that only 198 of these CSD partnerships were still active. The remaining 150 had either completed their activities or were otherwise no longer operational" (UNDESA, 2013, p. 5). But it is not clear how many of the remaining 150 had indeed satisfactorily achieved their objectives and how many were simply non-operational or when they became non-operational.

An additional point to consider is that the report noted that "the record showed that new CSD partnerships were not being registered" and further stated that while "stakeholders felt" that the:

> ...principle of partnership was still valid and was referred to several times as a concept in the Rio+20 outcome document, an alternative, more accessible and agile mechanism was needed to capture voluntary multi-stakeholder initiatives for sustainable development. This new mechanism would be the voluntary commitment, as described in paragraph 283 of the Rio+20 outcome document" (UNDESA, 2013, p. 5).

So now the distinction between PSDs and VC appears to be not just about PSD being related to social capital of associations whilst VCs are more implementation-related, but also that somehow VCs are an alternative, more accessible and agile mechanism for sustainable development action. But, an actual review of the FWW and of paragraph 283 which contains the sole reference to VCs does not reveal any clear description of what exactly constitutes a VC, and definitely does not mention that VCs are to be considered as somehow more accessible or agile a mechanism in delivering on sustainable development. In contrast to the FWW's limited mention of VCs, this UN report is quite emphatic in its statement that "voluntary commitment is distinguished from the negotiated commitments contained in the Rio+20 outcome document as well as other intergovernmentally agreed sustainable development outcomes".

The lack of clear definition regarding a VC within the FWW is troubling in terms of having clear globally agreed guidance on differentiating between voluntary initiatives. But what is harder to reconcile is the idea that is reflected in a UN report, but not in a globally agreed outcome document, that somehow PSDs are not to construed as implementation-focused and that the newest variant in the world of voluntary multi-stakeholder initiatives should be seen as more implementation-focused and responsive/agile. Given the recognition that no new partnerships were being registered with UNDESA, the question that needs to be carefully considered in the future is whether PSDs have fallen by the wayside to be superseded by VCs and, if this is the case, then why haven't VCs been better defined and how will existing PSDs be accounted for? The growth in numbers of registered voluntary initiatives that emanate from action networks without clear chronological markers makes it even more necessary that a comprehensive accountability framework be put in place. Overall, the lack of institutionally agreed guidance that would assist towards a comprehensive and globally verifiable framework for the implementation of the PSDs, the PPPs focused on sustainable development or the GPSD, and VCs remains a concern, especially in light of the growing importance of partnership activities in the UN-led post-2015 development agenda.

4.4   An abundance of voluntary initiatives but an absence of a universal accountability framework: Examining energy access and climate change initiatives

There is a plethora of partnerships and VCs, but a lack of institutional clarity in terms of how these various categories of voluntary efforts can be differentiated and accounted for. As discussed earlier, voluntary activities in the UN come in various shapes and forms and it is important to differentiate amongst the broad catch-call concept of voluntary multi-stakeholder efforts aimed at achieving sustainable development objectives.

Voluntary initiatives at the UN are not easy to decipher, because in addition to the UN-affiliated PSDs and VCs that have emanated as a result of the JPOI and the FWW outcomes, and which are contained in the SD in Action Registry, there are also partnerships affiliated with the UN Office of Partnerships (UNOP) that have particular resonance in terms of sustainable energy and energy access for the poor, and climate change issues. Then there is the issue of the partnerships listed with the UN Global Compact, which is listed as one of the action networks in the SD in Action Registry but which was actually launched in July 2000, and comprises "8,000 corporate participants in over 140 countries", representing "the world's largest voluntary corporate sustainability initiative" (UN Global Compact, 2014, p. 2).

Before discussing the issue of PSDs and VCs listed within the SD in Action Registry, it is useful to briefly point out that that partnerships were also the purview of the UNOP in the period 1998–2005 and have also been the focus of the UN Global Compact.

To be clear, the UNOP is not to be confused with UNOPS, which stands for the United Nations Office for Project Services and is charged with helping the UN and its partners run peace-building, humanitarian, and development operations. The stated aim of the UNOP, which served "as a gateway for partnership opportunities with the United Nations family" was to promote "new collaborations and alliances in furtherance of" the MDGs and to provide "support to new initiatives of the Secretary-General". In addition to providing Partnership Advisory Services and Outreach to a variety of entities, the UNOP managed the United Nations Fund for International Partnerships (UNFIP), established by the Secretary-General in March 1998 to serve as the interface in the partnership between the UN system and the UN Foundation (which was funded by a $ 1 billion pledge from Ted Turner in 1997), and the United Nations Democracy Fund (UNDEF), established by the Secretary-General in July 2005 to support democratization throughout the world (see the UNOP website: http://www.un.org/partnerships). The UNFIP has a number of identified priority areas where partnerships are focused: children's health, population and women, environment, peace and security and human rights, and financial contributions. What is interesting here is that, according to UNOP, the environment programme comprising 140 projects valued at $16.7 million focused on two thematic areas, biodiversity and sustainable energy and climate change, but the data and figures provided indicate that the program was operational from 1998 to 2005 (see the UNOP website; see also Box 4.1).

Partnerships and networks are also viewed as critical to the overall functioning of the UN's Global Compact – "a strategic policy initiative for businesses that are committed to aligning their operations and strategies with 10 universally accepted principles in areas of human rights, labour, environment, and anti-corruption" – and include environment-related principles 7–9: "support a precautionary approach to environmental challenges"; "undertake initiatives to promote greater environmental responsibility"; and "encourage the development and diffusion of environmentally friendly technologies" (UN Global Compact, 2014).

As of May 2, 2014, a search of the UN Global Compact Business Partnership Hub (https://businesspartnershiphub.org/) revealed the following action hubs:

- anti-corruption
- climate change and energy
- social enterprise
- UN business
- water.

Box 4.1   Partnerships/projects – United Nations Foundation/UN fund for International Partnerships (UNF/UNFIP) Sustainable Energy and Climate Change Program (1998–2005)

Renewable energy
- Rural Energy Enterprise Development (REED)
- Promoting Markets for Solar Water Heating in China
- India Solar Loan Facility

Energy efficiency
- Energy Efficiency Investment for Climate Change Mitigation
- Collaborative Labeling and Appliance Standards (CLASP) Programme

Promoting effective action on energy policy and climate change
- Energy Future Coalition (EFC)
- Institutional Investor Summit on Climate Risk
- International Climate Change Task Force

Source: UNOP website (http://www.un.org/partnerships).

The partnership hub is quite detailed in its structure because in addition to the five broad action hubs, it also includes additional search filters that allow searching by "action areas", "countries", "issue areas", "project" and "organizations". The filter of issues areas provides additionally detailed and specific sub-item searches; for instance, with regard to climate and energy, the sub-items include mitigation, adaptation, and finance.

A generalized search of the entire Business Partnership Hub with all the action hubs selected revealed a total of 312 current projects and organizations, but a more targeted and exclusive search of climate change and energy action hub revealed a total of 92 projects and organizations. When the filter of both "climate change and energy issues" was added to the climate change and energy action hub, a total of 71 projects and organization was listed, which indicated that the bulk of the projects and organizations in this action hub had a focus on climate change and energy, but the remaining projects and organizations, while affiliated with the action hub, may have other/additional issue areas of focus.

It is important to point out that a search of the action hub of climate change and energy does not allow for adding the search filter of sub-items, such as "improve/increase access to energy sources and "increase access to modern cooking appliances and fuels", "increase energy supply/distribution/use efficiency" and "produce/create financing solutions", that are listed under the action area of social enterprise – climate change and energy. That is to say, the climate change and energy action hub can only be cross-referenced and searched using the action area of climate change and energy, but not using energy access for the poor related filters. What is interesting is that when all the primary action hubs are selected along with the four sub-items given earlier, a total of 27 projects and organizations are listed, with only 10 being listed as actual projects relating to the four sub-items listed earlier. The idea that a search based on selecting the climate change and energy action hub does not allow one to input the sub-items listed above would appear to indicate a de-linkage in the partnership interface between energy access for the poor and climate change and energy.

The most definitive registry for partnerships is maintained by UNDESA as per the mandate of the FWW. According to the UNDESA, Rio+20:

> …was an action-oriented conference, where all stakeholders, including Major Groups, the UN System/IGOs, and Member States were invited to make commitments focusing on delivering concrete results for sustainable development on a voluntary basis. By the end of the Conference, over 700 voluntary commitments were announced and compiled into an online registry managed by the Rio+20 Secretariat, initiating a new bottom-up approach towards the advancement of sustainable development (UN Sustainable Development Knowledge Platform website, https://sustainablede-velopment.un.org/sdinaction/wssd).

Given the burgeoning growth of voluntary initiatives, the question is what exactly is the nature and scope of the UN registry that can allow for these voluntary initiatives to be clearly accessed and accounted for. Here, the need for a clear accountability framework is underscored because there appear to be some discrepancies in terms of how the UNDESA maintained SD in Action Registry accounts for PSDs and VCs and also some discrepancies between the Registry maintained by UNDESA and those listed in a 2013 UNDESA report.

A recent 2013 UN special report entitled *Voluntary Commitments & Partnerships for Sustainable Development* states that "voluntary commitments and partnerships for sustainable development are multi-stakeholder initiatives voluntarily undertaken by Governments, intergovernmental organizations, major groups and others that aim to contribute to the implementation of intergovernmentally agreed sustainable development goals and commitments" that are contained in 2012 FWW, Agenda 21, and the JPOI (UNDESA, 2013, p. 2). What is especially noteworthy about this report is that it happens to be the only "synthesis of current information on the 1,382 voluntary commitments, partnerships, initiatives and networks for sustainable development to date…"

Table 4.1   Breakdown of the 1,382 voluntary initiatives listed in UN Special Report, *Voluntary Commitments & Partnerships for Sustainable Development*

Voluntary initiatives available in the Sustainable Development in Action registry at the UN Sustainable Development Knowledge Platform	Number
Voluntary commitments individually registered	200
Sustainable Energy for All	120
UN Global Compact	125
Higher Education Sustainability Initiative	272
Green economy policies and practices	302
Sustainable Transport Action Network	21
World Summit on Sustainable Development (WSSD) Partnerships for Sustainable Development	198
Other Initiatives – Every Woman Every Child	144
Total	1,382

Source: UNDESA (2013, p. 6).

But there is an important caveat in terms of this report because the array of voluntary efforts are not all registered within the context of the WSSD and Rio+20 Conference registries, but also include efforts resulting from networks and other initiatives associated with SE4All, the United Nations Global Compact, the Higher Education Sustainability Initiative, green economy policies, practices and initiatives etc. Table 4.1 is excerpted from this 2013 synthesis report.

A May 2, 2014 search of the UNDESA's Sustainable Development Knowledge Platform's SD in Action registry includes a total of 1,427 initiatives (i.e. both PSDs from WSSD and PSDs and VCs from Rio+20). The SD in Action Registry is billed as a unified registry and contains two identical search tabs and filters. The first one is a left search tab by conference, namely "Partnerships for SD (CSD11/WSSD)" and "Rio+20". The second one is an advanced search filter also listing "Partnerships for SD (CSD 11/WSSD)" and "Rio+20", but this second one also includes a series of additional search filters on the following seven action networks:

- Every Woman Every Child
- Global Compact
- Green economy policies, practices and initiatives
- Higher Education Sustainability Initiative
- Sustainable Cities
- SE4All
- Sustainable Transport.

A search of the Registry reveals the number of PSDs related to CSD-11/WSSD as 196 rather than the 198 reflected in the UNDESA 2013 report. A search of the Registry using the filter of "Rio+20" lists 251 initiatives.

In the Sustainable Development Knowledge Platform, the partnerships registered in conjunction with the CSD 11-WSSD registry are listed alphabetically and are not clustered in thematic areas and contain information based on a template with two categories: "partners" and "description/achievement of the initiative". Partnership initiatives do not uniformly list start and end dates or resources committed and there are considerable variations in terms of description/achievement components across the initiatives.

The result included in Table 4.2 is based on applying the search filter of an individual action network and applying dual search filers of the action network plus the conference type – that is, "Partnerships for SD in CSD-11/WSSD" and "Rio +20". There are some important facts to keep in mind:

- The search indicates the total number of voluntary initiatives (PSDs and VCs) across both conferences as 1,427.
- The total number of voluntary initiatives listed under the CSD-11/WSSD category is 196 and the total number of voluntary initiatives (PSDs and VCs) listed under Rio+20 is 251, when the action network filters are not used.
- VCs are a new addition to the world of voluntary efforts related to sustainable development and came about only during Rio+20, whereas CSD-11/WSSD represents a static set of partnerships entered into a decade ago.
- The remaining subtotal of 1231 initiatives can therefore logically only be assigned to voluntary initiatives emerging from the action networks listed the SD in Action Registry, in the lead-up to and after Rio+20.

As Table 4.2 demonstrates, the numbers associated with individual action networks as of May 2, 2014 are very different from those contained in the 2013 UNDESA report. In addition, a search of the individual action networks and a search of individual action networks combined with the conference filters reveals discrepancies between the cumulative totals of voluntary initiatives associated with individual action networks and the totals reflected when the registry is searched as a whole based on search filters of all actions networks and conference filters. What this indicates is that VCs and PSDs associated with individual action networks are either being double counted or somehow cross-referenced across different action networks.

Based on the totals listed in the registry search, CSD-11/WSSD had 1,207 voluntary initiatives, including all action networks registered, and Rio+20 had 1,262, including all action networks. But the category of VCs was not in existence in the CSD-11/WSSD conference and action networks like SE4All had not even been launched at the time of WSSD. In fact, the figures for individual action networks reveal far more voluntary partnerships for SD

Table 4.2   Number of voluntary commitments (VCs) and partnerships for sustainable development (PSDs) based on search by action network and conference filters contained in the UN's Sustainable Development in Action Registry

VCs and partnerships in action network	VCs and PSDs in Action Network + partnerships for SD (CSD-11/WSSD)	VCs and PSDs in action network + Rio+20
Every Woman Every Child = 144	Every Woman Every Child = 340	Every Woman Every Child = 395
Global Compact = 124	Global Compact = 320	Global Compact = 375
Green Economy, Policies, practices, and initiatives = 319	Green economy, policies, practices, and initiatives = 515	Green economy, policies, practices, and initiatives = 570
Higher Education Sustainability Initiative = 284	Higher Education Sustainability Initiative = 480	Higher Education Sustainability Initiative = 535
Sustainable Cities = 8	Sustainable Cities = 204	Sustainable Cities = 259
SE4All = 119	SE4All = 315	SE4All = 370
Sustainable Transport = 13	Sustainable Transport = 209	Sustainable Transport = 264
Cumulative total = 1,011	Cumulative total = 2,383	Cumulative total = 2,768
Total reflected by registry search = 1,011	Total reflected by registry search = 1,207	Total reflected by registry search = 1,262

Source: UNDESA, Sustainable Development Knowledge Platform website (https://sustainabledevelopment.un.org/sdinaction/wssd).

related to CSD-11/WSSD in each of the individual action networks than were ever listed in the 2013 UNDESA synthesis report. And then there is the bigger challenge of explaining this much larger array of initiatives in light of the fact that the total number of PSDs listed under CSD-11/WSSD is only 196. The cumulative total of the VCs and PSDs associated with individual action networks for the CSD-11/WSSD simply do not add up, because clearly there are overlapping VCs and PSDs, but the question is, which ones are being double counted or referenced to more than one action network and why?

The difference in the number of PCs and VSDs associated with each action network and cross-tagged with a UN conference is hard explain, because it is not clear what such a search is supposed to reveal, especially as the actual number of VCs and PSDs associated with an individual action network is listed as being much smaller. It is also not clear how these individual PSDs and VCs are being tracked across conferences and action networks, or even whether they are being tracked within action networks across time. There are no apparent chronological search markers that can enable effective tracking of initiatives over time, other than the two search filters by conference. Adding

Table 4.3 Energy and climate change voluntary commitments (VCs) and partnerships for sustainable development (PSDs) contained in the UN's Sustainable Development in Action Registry

VCs and partnerships by topic filter	VCs and PSDs by topic and conference filter (partnerships for SD (CSD-11/WSSD)	VCs and PSDs by topic filter and conference (Rio+20)
Climate change = 31	Climate change = 5	Climate change = 14
Energy = 140	Energy = 28	Energy = 17
Poverty eradication = 3	Poverty eradication = 0	Poverty reduction = 2

Source: UNDESA, Sustainable Development Knowledge Platform website (https://sustainabledevelopment.un.org/sdinaction/wssd).

more confusion to the puzzle are the much more sobering numbers of voluntary initiatives associated with climate change and poverty eradication that are listed in separately from the SE4All action network filter. Table 4.3 provides a stark snapshot of the limited number of VCs and PSDs reflected in the Registry in relation to poverty eradication in particular, and to climate change.

Clearly, the issue of distinguishing between VC and PSD registries could benefit from a greater degree of institutional clarity and guidance. The overall discrepancies in clearly accounting for partnerships and VCs and accurately distinguishing between two categories of efforts needs to be urgently resolved and the only option is to have a transparent accountability framework. There is a need for institutional clarity in terms of distinguishing between VCs launched by action networks, such as those listed earlier, and individual PSDs and VCs, as well as the registries that would allow voluntary initiatives as a whole to be accounted for in terms of delivering on sustainable development and poverty eradication.

The FWW commits the UN system to put in place a detailed and systematic registry for VCs. According to the 2013 UNDESA report, the Rio+20 VCs "seek to be SMART – Specific, Measureable, Achievable, Resource based, and Time-bound – with great depth of information on the action plans of sustainable development implementation, which in turn has yielded a higher level of transparency and future accountability" (UNDESA, 2013, p. 5). Very shortly thereafter, the UNDESA report also noted that different registries, such as those maintained by the Natural Resources Defense Council (NRDC) and the Clinton Global Initiative (CGI), use different terms to describe voluntary initiatives for sustainable development – for example, the NRDC coined the term "PINCs" or "partnerships, initiatives, networks, and coalitions" and the CGI uses the term "commitments", but goes on to argue that these different terms "despite slight semantic variation, refer to the same class of voluntary initiatives for sustainable development" (UNDESA, 2013, p. 5).

The UNDESA report recognized the need for a "Voluntary Accountability Framework" and proposed that the Framework include:

1. The preparation of an annual special report of the *SD in Action* newsletter, summarizing progress in VCs and PSDs in the SD in Action registry, which would include the various action networks being able to "summarize achievements based on their individual reporting mechanisms and extensive dialogue with their constituency";
2. Maintenance of an up-to-date SD in Action registry "based on continuous input from stakeholders – thus keeping the SMART criteria – Specific, Measurable, Achievable, Resource-based, and Time-bound commitments and partnership";
3. Having third party independent reviews of commitments and partnerships "rather than solely relying on self-reporting" because "self-reporting tends to sometimes lean focus on achievements, which can compromise public perception of the integrity of the SD in Action Registry" (UNDESA, 2013, p. 31).

Semantics aside, the need for a clear definition and implementation reporting framework in terms of a systematic template that can allow for cross-time and cross-initiative comparisons and evaluations is critical to keep track of the vast array of voluntary initiatives. Clearly independent verification of voluntary initiatives is critical and should be built into any verification/accountability framework, but the idea that some PSDs and VCs can be summarized based on individual reporting methods related to the particular network with which they are affiliated appears to amount to an outsourcing of accountability for voluntary initiatives. The question that remains to be answered, therefore, is how to ensure a universally applicable and globally relevant accountability framework that will ensure transparency for all registered voluntary initiatives, including those that are not part of larger action or advocacy networks.

There should be little doubt that the stakes for implementation are clearly high based on estimated total value of all commitments, and that is exactly why an accountability framework is so important. An independent and joint review of the Rio+20 commitments prepared by the NRDC and the Stakeholder Forum pointed out that "of the 1,412 commitments registered in the UNDESA registry as of September 1, 2013, roughly 58 percent were made by the private sector and civil society, 30 percent from individual countries and 12 percent from the UN System", with the total estimated value of all commitments registered with UNDESA at "US\$637 billion" and "the monetary value of these commitments now amounts to 1 percent of yearly global GDP in total" (NRDC/Stakeholder Forum, 2013, p. 30). The NRDC reports also states that: "UNDESA does not currently have adequate resources devoted to the platform, resources that are necessary to build an architecture that would truly support, encourage, and hold commitment-makers accountable for the thousands of commitments made"(NRDC/ Stakeholder Forum, 2013, p. 30).

4.5 Need for improved clarity and accountability of voluntary initiatives on sustainable development

Voluntary efforts that are billed as non-substitutable, and complementary to intergovernmental action, are also specifically referenced as being critical to the implementation of UN globally agreed outcomes on sustainable development ranging from the 1992 Agenda 21 to the 2012 FWW. There is therefore an inherent dichotomy built into the wide variety of voluntary initiatives that have emerged over the course of the WSSD and the Rio+20 Summit, in that they are expected to assist with the implementation of sustainable development outcomes but are also expected to be complementary and additional to actual intergovernmental action. Voluntary partnerships and commitments are deemed non-binding, complementary, and non-substitutable for globally agreed goals, but are also seen as contributing directly to achievement of globally agreed goals, including the MDGs and presumably the final SDGs agreed in 2015. But how does the global community intend to track the progress of these voluntary efforts in the absence of a comprehensive and accountable registry and definitional clarity?

Differentiating amongst diverse voluntary efforts that have emerged as a direct response to key UN global sustainable outcomes is crucial. Meeting global challenges like increasing access to sustainable energy services for the poor and addressing climate change merit clearer operational and institutional guidance in terms of registering voluntary efforts. There is a need to pay focused attention to the role of voluntary initiatives, including those on energy and climate change, within the context of the UN, precisely because of the complicated relationship alluded to between voluntary efforts and globally agreed sustainable development-related outputs and goals.

The common feature that both voluntary partnerships and voluntary commitments for sustainable development share are that they do not commit relevant and involved actors to any legally binding action and are not envisaged as committing relevant stakeholders to any binding targets. Yet Agenda 21, the JPOI, and the FWW recognize the role of voluntary initiatives, including both partnerships and VCs, in implementing action related to sustainable development and the fulfillment of their respective globally agreed outcomes. It is this inherent tension between voluntary and largely self-reporting activities that are somehow supposed to be complementary to intergovernmental action but are also stated and seen as mechanisms by which implementation of agreed global outcomes on sustainable development can occur, that needs to be resolved. In fact, it is possible to argue that, given the absence of concrete intergovernmental action on energy access for the poor targets and goals, these voluntary initiatives become all the more important and have the real potential to play significant roles in the delivery of sustainable energy services to the poor, but then they all need clear and transparent accountability frameworks to ensure that they actually deliver on their objectives albeit voluntary and non-legally binding.

In terms of examining the two decades of key UN-led global outcomes on sustainable development, it is hard to understand why after Agenda 21 highlighted the need for a GPSD, it remained largely ignored until very recently. The somewhat disjointed trajectory of negotiated outcomes in terms of the GPSD is reflected in the fact that the WSSD's outcome, the JPOI, which pushed to prominence type 2 outcomes – PSDs – did not reference the GPD or the GPSD even once, and there happens to be only one reference contained in the FWW about the need to "reinvigorate" the GPSD after a 20-year dormancy. The ad hoc approach to the implementation of a robust global partnership in these key outcome documents does not point to any clear evolution of the concept.

According to the UN, PSDs in the context of WSSD and its follow-up are voluntary, multi-stakeholder initiatives intended to contribute to the implementation of sustainable development at various levels (national, regional, or global). In the decade that has ensued, PSDs have been officially recognized by the UN as a key mechanism for delivering on the promise of sustainable development, but there has been no consistent evaluation of the overall scope and track record of PSDs on energy for sustainable development, in particular energy access and climate change. The rapid growth of PSDs and VCs broadly focused on sustainable energy and climate change are not a substitute for intergovernmental action, but do serve as a contrast to the lack of intergovernmental negotiated agreements on concrete energy access for the poor and energy-related mitigation targets.

There is also a need for greater definitional and operational clarity in keeping track of voluntary initiatives, precisely because these initiatives have been growing in popularity and are supposed to be a key means of multi-stakeholder engagement as the UN pushes for a more inclusive and universal post-2015 development that was agreed to by the 2012 FWW outcome document. As a consequence of UN-led intergovernmental outcomes focused on sustainable development over the past 20 years, voluntary, multi-stakeholder initiatives that do not contribute to legally binding agreement are increasing in popularity and scope, but are also hard to comprehensively analyze, precisely because of the sheer range and diversity of the issues covered, the actors involved, and the outcomes anticipated. There appears to be an inherent dichotomy between the wide variety of multi-stakeholder-driven, voluntary efforts which are enshrined in the UN context as being complementary and non-substitutable for intergovernmental agreed actions, but which are also clearly referenced as being critical to the overall implementation of the 2012 FWW. There appears to be an overall institutionally derived challenge not only in differentiating amongst various kinds of voluntary initiatives, but, more importantly, in explaining how exactly these varied, voluntary and largely self-reporting efforts aimed at a wide variety of sustainable development issues can simultaneously be seen as complementary alternatives to globally agreed intergovernmental action, but also anticipated to play a vital role in the implementation of global agreed outcomes relating to sustainable development. The lack of endorsement of the SE4All initiative in the final FWW, in contrast to the recognition accorded to SE4All in other non-negotiated UN documents, has been pointed out. However, the recent adoption of the SE4All energy access goal in the 2014 OWG final outcome document as goal 7 is an indication that energy access for the poor is being actively championed.

In light of the recent agreement on SDGs and the anticipated targets associated with goal 7 (energy access) and goal 13 (climate change), it is crucial to have definitional clarity on differentiating between PSDs and VCs. Tracking potential progress toward SDG targets that are anticipated to be fulfilled by a combination of stakeholder actions, including VCs, requires clarity and accountability. Unfortunately, there does not appear to be a clearly stated definitional reference in the FWW, which is the institutional outcome document where VCs are first mentioned. There are also key questions that remain unanswered at an institutional and implementation level relating to the broad category of PSDs and VCs affiliated and/or registered with the UN. These questions include:

- Whether VCs can be devised as PSDs, and vice-versa, if so, what exactly are linkages and distinctions between these two categories of voluntary efforts?
- How are PSDs that could potentially contribute to VCs distinguished from partnerships that are convened under the aegis of the UN's Global Compact and UNOP?

Further complicating the understanding of voluntary efforts, including both PSDs and VCs, there is also the issue of how to understand the linkages between individual PSDs and VCs, and the broader role and function of the globally agreed goal of MDG-8 focused on developing a GPD, or the more amorphous goal of the GPSD, which is anticipated to have greater prominence in the post-2015 development agenda.

Having a concise, easily accessible yet transparent accountability framework of voluntary initiatives, particularly in the area of energy access for the poor and climate change, is important because these initiatives are increasingly being seen as playing an important role in assisting with the implementation of agreed global sustainable development outcomes.

The FWW mandated an accountability framework that is supposed to be transparent, accessible to the public and periodically updated, but key questions remain:

- What are the existing and agreed operational frameworks by which these varied, and largely self-reporting and non-binding voluntary efforts can be evaluated in terms of their contributions towards achieving the proposed SDGs?
- How exactly will the overarching GPSD and/or the GPD be linked to, or differentiated from, the broad array of individual partnership efforts aimed at specific sustainable development concerns in any post-2015 development agenda?

The provision of necessary levels of support and resources that can support a truly comprehensive, transparent accountability framework for all UN-registered voluntary initiatives on sustainable development, including those on energy and climate change, should be a global priority. The need for a coherent and transparent implementation framework for partnerships at the UN has been highlighted at the April 2014 Joint Thematic Debate/Forum on Partnerships, held jointly by the General Assembly and the Economic and Social Council, on "The role of partnerships in the implementation of the post-2015 development agenda". At this meeting the UN Secretary-General stressed that his proposal for a partnership facility which has been submitted to Member States for approval was meant to ensure that the organization had the capacity it needed to harness the strengths of its external partners. His point that "agreeing on an ambitious post-2015 development framework without preparing the world body to implement it would marginalize the institution at precisely the moment when it should be leading the charge" (UN Department of Public Information [UN DPI], 2014) can be seen as a clarion call for a more improved accountability framework for the diverse range of voluntary initiatives. In the final analysis, however, any accountability framework on voluntary initiatives will need the requisite amount of global political will to implement, because such a framework has the potential to reveal clearly what is being done, when, by whom and for whom, within what is currently a confusing mélange of collaborative activities.

References

Andonova, L. (2010) Public-private partnerships for the earth politics and patterns of hybrid authority in the multilateral system. *Global Environmental Politics*, 10(2), 25–53.

Colverson, S. and Perera, O. (2012) *Harnessing the Power of Public-Private Partnerships: The Role of Hybrid Financing Strategies in Sustainable Development.* Winnipeg: IISD. http://www.iisd.org/pdf/2012/harnessing_ppp.pdf [accessed on April 7, 2014].

Davidi, A. (2013) "Ban Ki-moon: why partnerships are key to fulfilling millennium goals". *The Guardian*. http://www.theguardian.com/media-network/media-network-blog/2013/nov/15/ban-ki-moon-millennium-development [accessed April 1, 2014].

Gemmill, B., Ivanova, M. and Ling, C-Y. (2002) Designing a new architecture for global environmental governance. *World Summit for Sustainable Development Briefing Paper*. London: International Institute for Environment and Development.

Glasbergen, P., Biermann, F. and Mol, A. (2008) *Partnerships, Governance and Sustainable Development: Reflections on Theory and Practice*. Northampton: Edward Elgar Publishing.

Glennie, J. (2014) Will competing UN and OECD partnerships stymie aid effectiveness? *The Guardian*, 11 April 2014. http://www.theguardian.com/global-development/poverty-matters/2014/apr/11/un-oecd-partnerships-aid-effectiveness [accessed April 29, 2014].

Government of India website. Public Private Partnerships. Available at http://www.pppinindia.com/ [accessed on April 7, 2014].

Joyner, C. (2005) Rethinking international environmental regimes: what role for partnership coalitions? *Journal of International Law & International Relations*, 1(1–2), 89–120.

Linder, S.H. and Vaillancourt, R.P. (2000) Mapping the terrain of the public-private policy partneship. In: Vaillancourt, R.P., ed. *Public-Private Policy Partnerships*. Cambridge MA: MIT Press, pp. 1–18.

NRDC/Stakeholder Forum (2013) *Fulfilling the Rio+20 Promises: Reviewing Progress since the UN Conference on Sustainable Development*. http://www.nrdc.org/international/rio_20/files/rio-20-report.pdf [accessed May 11, 2015].

OECD/UNDP (2014) *Making Development Co-operation More Effective: 2014 Progress Report*. Paris: OECD Publishing.

Pattberg, P., Biermann, F., Chander, S. and Mert, A. (2012) *Public-Private Partnerships for Sustainable Development: Emergence, Influence and Legitimacy*. Northampton: Edward Elgar Publishing.

Rehfuess, E. (2007) United Nations Commission on Sustainable Development – a missed opportunity for action on indoor air pollution? *Energy for Sustainable Development*, 11(2), 82–83.

Spalding-Fecher, R., Winkler, H. and Mwakasonda, S. (2005) Energy and the World Summit on Sustainable Development: what next? *Energy Policy*, 33(1), 99–112.

UN (1992) Agenda 21. http://sustainabledevelopment.un.org/content/documents/Agenda21.pdf [accessed February 25, 2014].

UN (2002) Plan of Implementation of the World Summit on Sustainable Development. Available at http://www.un.org/esa/sustdev/documents/WSSD_POI_PD/English/WSSD_PlanImpl.pdf [accessed April 12, 2014].

UN (2014) *The Global Partnership for Development: Making Rhetoric a Reality. MDG Task Force Report*. New York: UN Publications.

UN Commission on Sustainable Development (2002) *Guiding Principles for Partnerships for Sustainable Development ('type 2 outcomes') to be Elaborated by Interested Parties in the Context of the World Summit on Sustainable Development (WSSD)*. http://www.un.org/esa/sustdev/partnerships/guiding_principles7june2002.pdf [accessed March 27, 2014].

UNDESA (2013) Voluntary commitments & partnerships for sustainable Development. *Special Report of the SD in Action Newsletter*. http://sustainabledevelopment.un.org/content/documents/930Report%20on%20Voluntary%20Commitments%20and%20Partnerships.pdf [accessed April 9, 2014].

UNDESA website, Sustainable Development Knowledge Platform: Registry by Conferences and SD in Action Registry. http://sustainabledevelopment.un.org/index.php?menu=1348 [accessed on May 2, 2014.

UN DPI (2014) Secretary-General calls for Strengthening United Nations Capacity to Harness Strengths of External Actors in Tackling Development Goals . April 9, 2014. http://www.un.org/News/Press/docs/2014/ga11494.doc.htm

UNECOSOC (1993) *Commission on Sustainable Development: Report on the First Session*. E/1993/25/Add.1 E/CN.17/1993/3/Add.1. New York: UN.

UNECOSOC (1994) *Commission on Sustainable Development: Report on the Second Session*. E/1994/33 and E/CN.17/1994/20. New York: UN.

UNECOSOC (1995) *Commission on Sustainable Development: Report on the Third Session*. E/1995/32. New York: UN.

UNGA (2006) Resolution adopted by the General Assembly on 22 December 2005: Towards global partnerships. A/RES/60/215. New York: United Nations.

UNGA (2012) *The Future We Want*. Resolution adopted by the General Assembly on 27 July 2012. A/RES/66/288. New York: UN.

UNGA (2013) Outcome document of the special event to follow up efforts made towards achieving the Millennium Development Goals A/68/L4. New York: UN.

UN Global Compact (2014) *United Nations Global Compact: Corporate Sustainability in the World Economy*. New York: UN Global Compact Office.

UN Joint Inspection Unit (2013) *Review of the Management of Implementing Partners in United Nations Systems Organizations*. JIU/REP/2013/4. Geneva: UN. https://www.unjiu.org/en/reports-notes/JIU%20Products/JIU_REP_2013_4_Report%20and%20Annexes_English.pdf [accessed May 1, 2014].

UN Millennium Project (2010) *Goal 8: Develop a Global Partnership for Development*. Fact Sheet. New York: UN Millennium Project.

WHO (2007) *Indoor Air Pollution: Household Energy and Health at CSD-15*. http://www.who.int/indoorair/policy/hhhcsd15/en/ [accessed April 10, 2014].

5
Towards an Integrated Framework on Energy Access for the Poor and Climate Change
Issues to Consider for the UN-Led Post-2015 Development Agenda

The need for a single agenda is glaring, as soon as one starts thinking practically about what needs to be done. Right now, development, sustainable development and climate change are often seen as separate. They have separate mandates, separate financing streams, and separate processes for tracking progress and holding people accountable. This creates overlap and confusion when it comes to developing specific programs and projects on the ground. It is time to streamline the agenda.

UN (2013, p. 5)

In new estimates released today, WHO reports that in 2012 around 7 million people died – one in eight of total global deaths – as a result of air pollution exposure. This finding more than doubles previous estimates and confirms that air pollution is now the world's largest single environmental health risk. Reducing air pollution could save millions of lives.

WHO (2014a): http://www.who.int/mediacentre/news/releases/
2014/air-pollution/en/

Energy and Global Climate Change: Bridging the Sustainable Development Divide,
First Edition. Anilla Cherian.
© 2015 John Wiley & Sons, Ltd. Published 2015 by John Wiley & Sons, Ltd.

5.1   The absence of "energy access for poor" in the negotiating silos on climate change and sustainable development: Summary of findings

Securing a consensus-based comprehensive global framework agreement on climate change has become harder over the course of the past 5 years, when the bulk of countries across the globe have been dealing with the impacts of the global economic crisis, and the vagaries of global energy markets in relation to domestic energy security needs. But, as the chapters in this book demonstrate, ignoring the divide between intergovernmental negotiations on two pressing development challenges – global climate change and energy access for the poor – is wasteful and inhibits action that can address both challenges in a synergistic manner. The UN and its member states have spent a considerable amount of time and resources (financial, institutional, and human) on streamlining and integrating development efforts at the institutional and member state delivery level in order to reduce overlaps and improve effectiveness. It is perplexing, therefore, that the 2013 Report of the High level Panel of Eminent Persons on post-2015 development highlights the "glaring" need for a single agenda, one that avoids treating sustainable development and climate change as separate issues, and yet intergovernmental negotiations on these 2 issues continue to run on separate and different tracks. But the need for single agenda may be hard to achieve because separate, parallel negotiation silos on energy for sustainable development and climate change appear to be the long-standing and preferred mode of action within the context of the UN. Put simply, increasing access to sustainable energy services for the poor has never been effectively linked with global climate change objectives, nor until very recently has it even been reflected in terms of a concrete goal in any agreed UN outcome on sustainable development.

In spite of the expenditures of time, resources, and good intentions for promoting a more integrated UN-led sustainable development agenda, Chapters 2 and 3 provide evidence of distinctly separate negotiating silos on climate change and energy for sustainable development. These chapters reveal that the issue of energy access for the poor has consistently failed to secure high priority in either set of silo outcomes, and that it is only recently, in 2014, that energy access for the poor targets were agreed in the formulation of the sustainable development goals (SDGs). There is currently no integrated intergovernmental agreed framework for addressing two pressing and interlinked global development concerns that both fall under the broader rubric of sustainable development – namely increasing access to sustainable energy services for the poor and climate change.

The fact that there are two separate and parallel intergovernmental negotiations silos focused on addressing increasing access to energy services for the poor and global climate change concerns is hard to reconcile, precisely because two global development challenges are inherently linked in any UN-led quest for sustainable development. This "siloization" of intergovernmental negotiations on global climate change and energy for sustainable development (which is where energy access issues are discussed) needs to be urgently addressed because it is ineffective and inefficient given the UN's push for a shared and more inclusive post-2015 sustainable development agenda.

The synergies among energy, economic growth, and human welfare are arguably such that the overall challenge in finding a consensus framework agreement for greenhouse gas (GHG) mitigation encapsulating all major GHG-emitting countries has been precisely because any such global mitigation framework has been perceived as imposing limits on individual countries' energy needs and energy security, which, in turn, is viewed as impacting on national economic development. Clearly, the search for a global consensus that would respond to the effects of growing national energy demands, particularly by large developing countries seeking to address both socio-economic development and poverty reduction objectives, and mitigate against increasing GHG emissions, in the face of a vast array of development needs and risks, is a complex task. But, as Chapters 2 and 3 demonstrate, ignoring the chasm or divide between two sets of intergovernmental negotiation silos on what are arguably two of the most pressing development challenges – global climate change and energy access for poverty reduction – is both wasteful and ineffective. Meanwhile, the vibrant growth in voluntary initiatives related to sustainable development, and the growing role of such initiatives in addressing sustainable development concerns in the absence of definitional clarity, and a robust accountability framework for voluntary initiatives within the UN context pose serious challenges for accessing the efficacy and delivery of results emanating from these voluntary initiatives.

The examination of key UN-led global outcomes on sustainable development and climate change reveals that, for decades now, the two agendas have been consistently separated out with few consistent attempts at programmatic and policy linkages between energy for sustainable development and climate change objectives. As Chapter 4 demonstrates, the increasing importance of voluntary initiatives for sustainable development, including energy access and climate change-related initiatives, could benefit from improved transparency and accountability frameworks. Simply separating out the programmatic and policy linkages between energy access for the poor and climate change mitigation perpetuates the idea that energy for sustainable development and climate change concerns should somehow continue to be considered as separate silos within the UN context. It is time to take stock that these separate negotiating silos have been known to exist despite global rhetoric and evidence that they represent dual and intersecting global concerns, and that a shared and universal post-2015 development agenda will be much harder to accomplish if these silos persist.

As the previous chapters evidenced, there are numerous globally relevant scientific assessments pointing to the role of energy as a prime driver of anthropogenic global climate change, and many noteworthy UN reports highlighting the negative and serious impacts of energy poverty (the lack of access to sustainable energy services) on the lives of the poor. In addition, there is the global recognition that the poor have the least resilience and greatest vulnerabilities to the adverse climatic impacts. Notwithstanding a surfeit of global assessments and UN reports on the importance of energy as a key factor in global climate change and as a key input in reducing poverty, which happens to be a central tenet in the development agenda of the UN, a review of key intergovernmental outcomes on sustainable development and climate change indicates that programmatic guidance and recommendations on energy access for the poor have been largely been absent in the key globally agreed outcomes surveyed.

Based on close examination of the actual record of globally negotiated outcomes in climate change over the past two decades, and sustainable development over the past four decades, it is possible to summarize the following findings:

- Increasing access to sustainable energy services for the poor in developing countries has not been accorded global priority in either of the negotiations tracks and outcomes, and to date, there is an absence of concrete global targets relating to energy access for the poor.
- At the intergovernmental level, despite the rhetoric and the decades of resources expended, the two global challenges of poverty reduction and climate change are seen as falling under the purview of separate negotiations tracks and treated as distinct negotiated silos rather than intersecting and linked sustainable development related challenges.
- Energy for sustainable development, including energy access for the poor, and climate change mitigation objectives are treated as separate categories even within the same global negotiated outcome in key UN global outcomes related to sustainable development.
- Key outcomes of the intergovernmental negotiations on climate change, ranging from the 1992 United Nations Framework Convention on Climate Change (UNFCCC) to the 2013 Warsaw Outcomes, have not focused on providing concrete policy guidance and programmatic linkages related to energy for sustainable development in general and energy access for the poor in particular.
- Key global agreed outcomes on sustainable development reveal that the programmatic and policy nexus between energy, poverty reduction, and climate change is largely disjointed and that concrete references to energy access for the poor have been diminishing rather than increasing with the passage of time.

- While there has been tremendous growth in terms of the range of voluntary initiatives focused on sustainable development, the lack of an integrated framework that can simultaneously address energy for sustainable development concerns, including energy access for the poor and global climate change, inhibits engagement by multi-stakeholders interested in addressing both challenges in an integrated manner.

The absence of a policy and programmatic nexus between climate change and energy access for the poor at the global level can be traced to the fact that global negotiations on climate change conducted under the aegis of the UNFCCC have remained distinct from UN-led intergovernmental discussions on sustainable development. Global climate change negotiations are considered legally binding, whereas the broader UN global negotiations on sustainable development touch on a wider range of issues, including climate change, but do not comprise legally binding agreements and protocols.

Intergovernmental policy discussions and negotiations related to the newly emerging SDGs, and recent global initiatives shepherded by the UN, such as the Sustainable Energy for All (SE4All) initiatives, which target energy access, energy efficiency, and renewable energy objectives, therefore remain distinct from the intergovernmental climate change negotiations arena and any related discussions on nationally appropriate mitigation actions (NAMAs) and/or scaling up of levels of ambition to address the Durban Platform. Having separate intergovernmental negotiations on two global policy challenges that inherently lend themselves to synergistic action arguably inhibits innovation and cooperation, and is an unfortunate drain on a limited financial resources and technical capacities.

5.2   Separate silos pose challenges for a shared post-2015 development agenda

Given all the factors listed in the previous section, it is hard to understand how two separate silos on energy for sustainable development and global climate change within the UN context have actually been beneficial or will be effective in terms of a post-2015 development agenda? As a direct result of the 2012 *Future We Want* (FWW; United Nations General Assembly [UNGA], 2012) outcome, the UN has embarked on a quest for a more universal post-2015 development agenda, and this in turn provides an impetus for a more integrated intergovernmental framework that links energy for sustainable development, in particular energy access for the poor, with climate change concerns.

The looming 2015 deadline provides a stark reality check for the global community which seeks, simultaneously but separately, to secure a comprehensive global climate change agreement, and an agreement on a shared development agenda comprising of new set of SDGs, including targets on energy access and climate change. With 2015 fast approaching, there has been little integration between energy for sustainable development and a climate change negotiated outcome, and there is also the matter of considering just how arduous and slow-moving the UN-led intergovernmental process on energy-related climate change mitigation has become.

The clear evocation of the precautionary principle within the UNFCCC, and the central role of energy in driving climate change mitigation should have bolstered and spurred intergovernmental action to undertake energy-related mitigation, but the record of the climate change negotiations tells a different story. With 20 annual cycles of intergovernmental negotiations culminating in yearly Conferences of the Parties (COPs) occurring in diverse corners of the globe since the adoption of the historic 1992 UNFCCC, it is time to stop ignoring the role of energy as a chasm between developed and developing, oil-producing and oil-importing, coal-exporting and coal-importing, energy-secure and heavily energy-dependent countries, which are as evident today, if not more so, as they were 25 years ago.

The crucial role of energy in mitigating anthropogenic climate change and in poverty reduction, as well as the shared 2015 deadline for securing UN-led global agreement on these dual challenges, makes ignoring the linkages between sustainable energy access for the poor and climate change increasingly unjustifiable within the UN context. The proverbial bottom line is that both sets of intergovernmental negotiations have not yielded tangible results in the form of either a comprehensive GHG mitigation agreement that commits all large-scale aggregate emitters, or globally agreed targets and clear programmatic guidance on increasing access to sustainable energy services for the poor. After decades of intergovernmental negotiations on climate change and sustainable development, with their marked proclivity towards parsing negotiating language and convening all-night meetings in diverse cities, key globally agreed outcomes from these two non-intersecting negotiations silos have very little to show in terms of concrete energy-related mitigation and/or energy access targets and goals.

In a bid to ramp up political commitment for tackling global climate change, the UN Secretary-General called on "leaders of government, business, finance and civil society to bring bold announcements and actions to address climate change" within the context of a Summit to be held on September 23, 2014 (press release; UN, 2014a). The Secretary-General specifically called on China to provide "global leadership" in responding to climate change (Yeo, 2014). The outcomes and announcements emanating from the Climate Summit envisaged as a "solutions-focused Summit" convened in the UN headquarters a day before the formal opening of the UN General Assembly are seen as essential in contributing towards high-level political momentum for addressing climate change, but complementary to the formal climate change intergovernmental negotiations process. In other words, the UN is clear to point out that the Summit "is not another negotiation track, and there will be no negotiated outcome" (press release, UN, 2014b), but instead a Chair's summary will capture the Summit's outcome. The clear distinction between the non-negotiated outcomes and announcements made at the 2014 Summit, and negotiated global outcomes that will emerge within the context of the UNFCCC-led COP-20 and COP-21 negotiations, is important to reflect on. The exact modalities by which the "bold" but non-negotiated announcements made at a non-negotiated forum will be translated into the separately convened UNFCCC intergovernmental process remains to be seen.

"Bold and new" announcements by leaders of key aggregate emitters were invited and anticipated, but with leaders from the key countries China, India, Germany, and Australia not in attendance, expectations were dampened. In the end, what was said by three largest aggregate GHG-emitting countries is worth considering briefly.

The official statement made by China, the world's largest aggregate GHG emitter, was delivered by Vice Premier Zhang Gaoli, the Special Envoy of the Chinese President Xi Jinping who did not attend the Summit. Gaoli began by saying that "China is ready to work with the international community to actively tackle the grave challenge of climate change." He went on to point out that "China attaches high importance to addressing climate change" and highlighted the fact that China was the first among developing countries to formulate and implement a national climate change program and had "recently adopted a national plan on climate change to meet the target of cutting carbon intensity by 40 to 45 percent by 2020 from the 2005 level". Mirroring the call made by Obama about the role of "big" countries, Gaoli also noted that: "As a responsible major country, China will make greater effort to more effectively address climate change and take on international responsibilities that are commensurate with its national conditions, stage of development and actual capabilities". He added that China would undertake post-2020 actions on climate change as soon as it could, which would "include reducing carbon intensity and peaking of total CO_2 emissions as early as possible". As discussed in Chapter 1, there can be no mistaking that China views the energy sector as a key driver in addressing global climate change. While there was no mention of the term poverty or poor in the statement, Gaoli clearly signaled China's support for South-South cooperation to address climate change (UN Climate Summit 2014/ Executive Office of the UN Secretary-General, 2014).

The official statement made by the US – the world's second largest aggregate GHG emitter – was delivered by President Obama. The statement did not include any new emission target-related announcement. Obama did, however, state that the US "would meet its target of reducing carbon emissions in the range of 17 percent below 2005 levels by the year 2020" and that by early 2015, the US would "put forward" its "next emission target". Obama indicated that he had met with the Special Envoy of China to "reiterate" his belief that "as the two largest economies and emitters in the world, we have a special responsibility to lead. That's what big nations have to do". He went on to say:

> We recognise our role in creating this problem. We embrace our responsibility to combat it. We will do our part and we will help developing nations to do theirs. But we can only succeed in combating climate change if we are joined in this effort by every nation – developed and developing alike. Nobody gets a pass. (Whitehouse Office of the Press Secretary, 2014)

The focus of the latter part of his speech at the Summit was to call attention to the role of emerging economics that are "likely to produce more and more carbon emissions in the years to come" and to once again point out that no country could avoid getting involved or "stand on the sidelines on this issue".

India, the world's third largest aggregate GHG emitter, was also not represented at the highest level at the Summit. Speaking on behalf of the Prime Minister Modi, the Minister of Environment, Forests and Climate Change highlighted a different facet of the climate change crisis by pointing up front to the role of "poverty as a polluter" and to the UN statistics indicating that 1.2 billion people were still living in poverty. He went on to state that the current per-capita energy consumption in India is a fraction of figures for the developed world and stated that "energy consumption in India would need to increase by 4 times as India's Human Development Index increases" (Ministry of Information and Broadcasting, 2014). What this would require according to him was a "coupling of higher energy consumption with lower carbon intensity" and accordingly India was "fully committed" to achieving its "voluntary goal for reducing emission intensity of its GDP by 20–25% by 2020 over 2005 level". New developments in renewables and energy efficiency made by India were cited. With regard to energy access for the poor, the Minister noted that India was "committed to pursuing a path of sustainable development through eradication of poverty both of income as well as energy" and called on additional finance and technology support for developing countries . The interesting contrast is that, unlike China, India did not mention that it would be ready to take on international responsibilities to address climate change. While there was mention of reducing emission intensity and numerous references to sectoral energy efficiency measures being put in place, there was no mention of targets or measures related to increasing access to sustainable energy services for the poor.

The overall assumption is that the statements made at the Summit about ramping up climate change mitigation will translate into concrete negotiated results in the separately convened UNFCCC COP-20 in order to enable a consensus-based global climate agreement by COP-21 in 2015. The long history of parsing negotiation language within the UNFCCC context, as well as the distinction between negotiated and non-negotiated outcomes, make for uncertainty as to what extent the "bold" non-negotiated outcomes from the 2014 Climate Summit will be reflected in the UNFCCC negotiating process that is expected to culminate in a legally binding global climate change agreement in the 2015 COP-21.

Another factor that is worth highlighting is that as a direct outcome of the 2012 Rio+20 Conference, the previous negotiating forum – the UN Commission on Sustainable Development (CSD), where energy for sustainable development issues were considered since the 1992 Rio Summit – has now been replaced by the High Level Political Forum (HLPF). The anticipated objective of the HLPF, which is a joint endeavor between the UNGA and the United Nations Economic and Social Council (UNECOSOC), is to promote and implement an integrated and balanced sustainable development agenda beyond 2015. It is intended as a universal platform on issues related to the implementation of sustainable development, including energy, and is supposed to pull together relevant actors on the global stage and within the United Nations system, and result in a level of coordination, integration, and coherence that until now has been lacking. The question that remains is how and whether it will tackle energy and climate change in a linked manner.

Figure 5.1 provides a view of the different UN processes focused on securing agreement on a post-2015 development agenda and climate change by 2015. As discussed previously, the recent OWG outcome document on SDGs revealed that any climate change related target will be separately arrived at under the aegis of the UNFCCC and then relayed into the UN GA final agreement on SDGs and the post-2015 development agenda.

As the record of 20 years of global climate change outcomes revealed, securing a consensus-based agreement on climate change mitigation is fraught with great difficulty and political tension, and a detailed programmatic focus on energy, in particular energy services for the poor has been surprisingly absent. It is time to acknowledge that global agreement for a timely and comprehensive resolution to the climate change challenge has become harder, not easier, over the course of the past 5 years, when the bulk of countries across the globe have been dealing the impacts of the global economic crisis and have been burdened with escalating energy costs and growing energy security concerns. But it is important to point out that these political and economic realities are much harder, and the costs and burdens of responding to climate change are much more severe for developing countries as they juggle development-related energy security and energy access needs along with global climate change mitigation and adaptation concerns.

While there is an active process of negotiations being undertaken with the context of the UNGA and in the Open-ended Group on SDGs, it remains

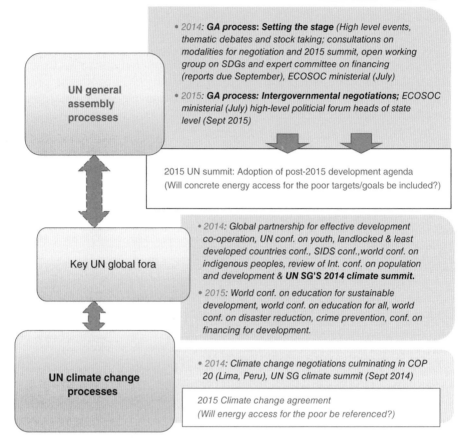

Notes:

(i) There are two separate UN-led processes for any future post-2015 development agenda and climate change agreements with different end dates for adoption in 2015.

(ii) The broken arrows represent potential for feedback and carryover of agreed texts by member states from key global fora into UN General Assembly and climate change processes, respectively.

(iii) UN-led climate change negotiations are convened separately from General Assembly processes related to post-2015 development agenda, as well as key UN fora.

(iv) It is unclear at this point whether concrete energy access targets for the poor and energy sector mitigation targets will be specifically referenced.

Figure 5.1 Key UN processes focused on a universal post-2015 development agenda and a comprehensive climate change agreement (2014–2015): separate silos.

to be seen whether and how climate change will factor into the SDGs. The 2014 agreed outcome on SDGs indicates that any climate change-related agreement will need to be officially arrived at within the separate context of the UNFCCC – that is, outside of the UNGA process given the legal primacy accorded to the UNFCCC negotiations. COP-21 is anticipated to arrive at a comprehensive climate agreement in December 2015, but the broad elements of this agreement will need to be agreed to in COP-20 in Lima, Peru, by 2014. However, the convening of the HLPF at the heads of state level in September 2015 is anticipated to forecast the adoption of the post-2015 development agenda, but this meeting will occur before any potential climate agreement has been arrived at. The different end dates for the adoption of two different global outcomes documents in 2015 poses a challenge for an integrated and mainstream development agenda. Ideally, there should be agreement on climate change which could be factored and referenced into the broader UN post-2015 development agenda. At this point in time, there is a placeholder SDG on climate change that is pending the final agreement in COP-21 and will only then be factored into post-2015 development agenda.

The coinciding of the 2015 deadline for a new climate change agreement, the achievement of the Millennium Development Goals (MDGs), and the adoption of a shared post-2105 development agenda make for a very unique opportunity to focus on synergies that allow for scaling up action related to an integrated development agenda. The UN's recent move towards formulating a new, streamlined and universal post-2015 development agenda with a set of shared SDGs cannot be effectively put into place within a global context in which separate negotiated silos and outcomes on energy, climate change, and poverty reduction continue to exist.

The need for a mainstream UN development agenda is also important as a frame of reference guiding the inputs and efforts of a wide variety of voluntary initiatives that have blossomed over the past decade. What is also interesting to note is that the absence of an integrated globally negotiated framework for addressing climate change and energy access should be contrasted with the rapid growth in new forms of partnerships and voluntary initiatives in these areas. As the discussion in Chapter 4 pointed out, the exponential growth of voluntary initiatives, including voluntary partnerships and commitments, are increasingly heralded as mechanisms for delivering on sustainable development within the UN global context. These voluntary initiatives are promising and diverse in application and scope and therefore also in need of a clearer accountability framework.

The idea that a framework of alliances, networks, and partnerships could work in tandem to deliver on the programmatic and policy nexus between energy access for the poor and climate change is borne out by diverse collaborative initiatives, some of which are discussed in the Section 5.3. The question that faces the global development community as it pursues a universal post-2015 development agenda is whether it can continue the trajectory of consigning close to 2.4 billion people to continue their reliance on inefficient and polluting biomass fuels (wood, dung, crop residues) to meet their cooking needs and heat their homes. Heavy dependence on the use of solid fuels in open fires and rudimentary stoves is not only inefficient, but can also result in high levels of indoor air pollution and emissions of short-lived climate pollutants (SLCPs) and impacts most severely on the lives of women and girls – who usually bear the burden of collecting the fuel and pay the price of being close to the polluting fuels sources. It is in this regard that it is important to consider new non-UN-led initiatives and alliances that aim to focus on the nexus between energy access and climate change, including the linkages between improving access to sustainable energy services and reducing emissions of SCLPs in poor households and communities.

5.3   Increasing energy access for the poor and reducing SLCPs: Two key global multi-stakeholder initiatives

The central role of energy security in the needs of all countries across the world cannot ever be discounted and it would be naïve to assume that concrete linkages between energy for sustainable development and climate change mitigation will be arrived at either easily or quickly. The absence of clear guidance and references linking energy for sustainable development with climate change within the key global outcomes is an indicator of just how politically contested energy as a topic is within both sets of negotiations. An integrated development agenda that can provide programmatic guidance and implementation on energy access for the poor and climate change is not going to be politically easy to agree upon, because energy for sustainable development issues, including energy access, has been heavily contested at the CSD and the 2012 Rio+20 Summit. But the scope and urgency of increasing access to sustainable energy services for the poor, whilst mitigating SLCPs, are such that scaled-up action in both UN-led intergovernmental and multi-stakeholder-driven arenas is needed to work in effective tandem to address these intersecting challenges.

Although the nexus between energy access for the poor and climate change has not been adequately addressed in an integrated manner with the context of key globally negotiated outcomes, there have been some key multi-stakeholder initiatives emanating from within and outside of the UN that have aimed to address this nexus in different ways. The most prominent of these from the UN's perspective has been the SE4All initiative, which is discussed in Chapters 3 and 4, but there are others, such as the Major Economies Forum, that constitute a break from the traditional intergovernmental forum by being a forum solely for major GHG emitters. The point is not to evaluate the merits or demerits of either intergovernmental or more issue-based multi-stakeholder initiatives, because clearly both forms of action are needed and should be supported. Intergovernmental and civil society actors are both needed to focus on the challenge of how to deal with separate silos and separate outcomes on what are intrinsically linked development challenges.

An integrated development agenda that can deliver on increasing access to sustainable energy services for the poor is ill-served by drawing distinctions between options that can work within the UN and outside of the UN, because both globally negotiated options and a diverse set of voluntary initiatives will be needed. As Chapters 3 and 4 demonstrate, key globally negotiated outcomes on sustainable development have pointed to an active engagement and collaboration between governments and civil society partners and ascribe a key role to voluntary initiatives in supporting the achievement of agreed global goals. The issue of whether or not global climate change and poverty reduction can be addressed via a web of diverse voluntary initiatives that can scaffold globally agreed climate change and poverty reduction goals is no longer the subject of expert debate but the realm of practical action.

There are two key multilateral and global coalitions that are focused directly on increasing access to clean energy services and reduction of polluting emissions which are worth briefly touching upon. Initially formulated as type 2 partnership, the Partnership for Clean Indoor Air (PCIA), which has been integrated into the Global Alliance on Clean Cookstoves (GACC), and the much more recent 2012 Climate & Clean Air Coalition to Reduce Short-Lived Climate Pollutants (CCAC) are discussed below. What is also interesting to consider is the possibility of a broader and more comprehensive partnership, given the close operational linkages between the CCAC and the GACC, and the fact that the CCAC seeks to build on catalyzing the existing efforts of the GACC. These three well-known initiatives are briefly discussed in the following because they focus on the real potential to address the nexus between climate change and energy access for poor that the more institutionalized UN-led processes related to energy for sustainable development and climate change have not been able to consider in great detail. However, the information contained below is derived primarily from the initiatives themselves and therefore is not independently verified but is intended to provide an informational snapshot of key multi-stakeholder initiatives that could offer fora for scaled-up action on reducing SLCPs whilst also increasing sustainable energy access for the poor.

PCIA and GACC According to the legacy website of the PCIA (http://www.pciaonline.org/), this partnership was originally formulated by the US Environmental Protection Agency and debuted on the global stage at the 2002 World Summit on Sustainable Development (WSSD) in Johannesburg. The PCIA was integrated with the GACC, but the resources that were produced over the past 10 years of the PCIA are still accessible on the legacy website. From its inception in 2002 until its integration into the GACC, 590 partner organizations joined together through the PCIA to contribute their resources and expertise to reduce smoke exposure from cooking and heating practices in households around the world. The PCIA was a "collaborative effort" focused on four elements:

- Meeting the needs of local communities for clean, efficient, affordable and safe cooking and heating options;
- Improving cooking technologies, fuels and practices for reducing indoor air pollution;
- Developing commercial markets for clean and efficient technologies and fuels;
- Monitoring and evaluating the health, social, economic and environmental impact of household energy interventions" (http://www.pciaonline.org/).

In September 2010, the US Department of State and the Environmental Protection Agency helped launch the GACC at the Clinton Global Initiative annual meeting in New York, NY. The Alliance is a public–private partnership run by a Secretariat hosted by the UN Foundation that "works closely with the more than 350 international partners from the private, government, UN, non-governmental, and academic sectors". The mission of the GACC is "to mobilize high-level national and donor commitments toward the goal of universal adoption of clean cookstoves and fuels" and more specifically "to foster the adoption of clean cook-stoves and fuels in 100 million households by 2020". GACC has "adopted a three-pronged strategy – enhance demand, strengthen supply and foster an enabling environment" to promote its goal, and has developed a 10-year business plan comprising: "Phase One (2012–2014): Launch global and in-country efforts to rapidly grow the sector, Phase Two (2015–2017): Drive investments, innovation and operations to scale; and Phase Three (2018–2020): Establish thriving and sustainable global market for clean cookstoves". In order to reach its goal, the Alliance is looking to attract £1 billion in investment in the sector by 2020. While the funding amounts that have been raised are not available on the GACC's website, the GACC is structured to receive funds from four broad donor partners- national, corporate, individuals and founda-tions and civil society (see the GACC website: http://www.cleancookstoves.org/the-alliance/).

CCAC In February 2012, the governments of Bangladesh, Canada, Ghana, Mexico, Sweden and the United States (six countries), along with the United Nations Environment Programme (UNEP), created the CCAC – a voluntary, and collaborative global coalition "to support fast action" related to SLCPs. According to the CCAC website (www.unep.org/ccac), "recognizing that mitigation of the impacts of short-lived climate pollutants is critical in the near term for addressing climate change and that there are many cost-effective options available", these six countries along with UNEP initiated "the first effort to treat these pollutants as a collective challenge" because "reducing them will protect human health and the environment now and slow the rate of climate change within the first half of this century".

The governance structure of the CCAC includes the following: a Working Group (with representatives from the partners, oversees the cooperative actions of the Coalition); a High-Level Assembly of the Coalition partners (which convenes to set policy, take stock of progress, and initiate future efforts); a Scientific Advisory Panel, which is responsible for keeping the Coalition abreast of new science development on SLCPs, answer specific questions of the Coalition and inform policy discussions; and a Secretariat, hosted by the UNEP in Paris. Interestingly, the CCAC was devised as a catalyzing partnership that would build on the existing work of GACC, the Arctic Council, the Montreal Protocol, and the Global Methane Initiative (GMI) and was intended to add to and complement, not replace or substitute for, global action aimed reducing CO_2 (www.unep.org/ccac).

In announcing the 2012 launch of the CCAC, the US State Department pointed out that:

> Fast action to reduce short-lived climate pollutants can have a direct impact on global warming, with the potential to reduce the warming expected by 2050 by as much as 0.5 Celsius degrees. At the same time, by 2030, such action can prevent millions of premature deaths, while also avoiding the annual loss of more than 30 million tons of crops. Moreover, many of these benefits can be achieved at low cost and with significant energy savings. (US State Department, 2012)

According to its March 2014 Executive Summary, the CCAC has grown "ten fold" since its inception 2 years ago, and currently has 87 partners, including 38 state partners including the European Commission; and 48 non-state partners including 12 intergovernmental organizations and 37 non-governmental organizations (NGOs), "which have endorsed the Framework for the Coalition and agreed to take meaningful action to reduce SLCPs" (CCAC, 2014, p. 8). The CCAC has 10 high impact initiatives, including one entitled, "Reducing SLCPs from household cooking and domestic heating" which aims to "speed up reductions in SLCP emissions through high-level advocacy, support for new finance mechanisms, new research, and development of standards and testing protocols to provide clear criteria for evaluating emissions reductions for improved cookstoves, heatstoves and fuels" (CCAC, 2014). Amongst the lead partners listed are Nigeria and the GACC. Out of the total of 10 initiatives, seven are listed as sectoral initiatives (CCAC, 2014, pp. 3–7):

1. Accelerating methane and black carbon reductions from oil and natural gas production;
2. Addressing SLCPs from agriculture;

3. Mitigating SLCPs and other pollutants from brick production;
4. Mitigating SLCPs from municipal solid waste;
5. Promoting hydrofluorocarbon (HFC) alternative technology and standards;
6. Reducing black carbon emissions from heavy-duty diesel vehicles and engines;
7. Reducing SLCPs from household cooking and domestic heating.

The remaining three are cross-cutting initiatives:

8. Financing mitigation of SLCPs;
9. Regional assessments of SLCPs;
10. Supporting national planning for action on SLCPs initiative (SNAP).

5.4   Bold action is needed to address the nexus between energy access for the poor and climate change: Possibilities for consideration

Energy sector issues lie at the heart of any resolution of the global climate change problem and are also critical for any shared post-2015 development agenda that puts poverty reduction at its core. Having two separate intergovernmental negotiating tracks on energy for sustainable development and global climate change with no intersecting or linking mechanisms prevents broader-scale engagement on energy, including reducing SLCPs by improving access to more sustainable energy services. There has been a rapid proliferation of complex negotiating groups within the context of the UNFCCC, but an evaluation of their outputs indicates that specific references to viable and practical energy sector options related to the poor are seldom, if at all, mentioned.

It is time for bold global action on the neglected yet crucial policy linkages between energy for sustainable development (services, systems and technologies) and global climate change, because these linkages hold the key to a broader and diverse range of national engagement in any future global climate change agreement as well as any post-2015 development agenda.

Undertaking "bold action" on global climate change by neglecting the nexus between climate change and increasing access to sustainable and cost-effective energy services for the poor is possible, but harder to justify and rationalize in the context of a UN push for a shared post-2015 development agenda. The impetus for "bold action" within the UN context lies in addressing energy access for the poor and climate change, not as separately negotiated silos, but in an integrated manner, because addressing global climate change mitigation, including SLCPs, should be seen as a critical element of the broader post-2015 UN global sustainable development agenda. Any call for "bold action" that integrates climate change and energy access for the poor objectives cannot be seen as a means for delaying requisite, nationally driven and/or globally appropriate actions taken to mitigate anthropogenic GHGs, but instead as means to allow for more effective, scaled-up action.

Bold action, in fact, necessitates that both UN and non-UN forums and groups work actively to promote an integrated agenda that can link practical measures for increasing energy access for the poor with climate change objectives. One such non-UN forum is the Major Economies Forum on Energy and Climate (MEF), which was explicitly established in recognition of the direct relationship between energy and climate change. Originally proposed as a US initiative, MEF comprised 17 major GHG emitters, including both developed and developing countries – Australia, Brazil, Canada, China, France, Germany, India, Indonesia, Italy, Japan, the Republic of Korea, Mexico, Russia, South Africa, the United Kingdom, and the United States – as well as the European Union, who met L'Aquila, Italy, on July 9, 2009. The first declaration of the MEF stated:

> Climate change is one of the greatest challenges of our time. As leaders of the world's major economies, both developed and developing, we intend to respond vigorously to this challenge, being convinced that climate change poses a clear danger requiring an extraordinary global response, that the response should respect the priority of economic and social development of developing countries, that moving to a low-carbon economy is an opportunity to promote continued economic growth and sustainable development, that the need for and deployment of transformational clean energy technologies at lowest possible cost are urgent, and that the response must involve balanced attention to mitigation and adaptation. (MEF, 2009)

The MEF has been billed as a forum intended to promote:

> candid dialogue among major developed and developing economies, help generate the political leadership necessary to achieve a successful outcome at the annual UN climate negotiations, and advance the exploration of concrete initiatives and joint ventures that increase the supply of clean energy while cutting greenhouse gas emissions. (MEF website, http://www.majoreconomiesforum.org/)

Given that the MEF comprises the world's major GHG emitters, and that it is envisaged as a forum to help, amongst other things, to augment political leadership for achieving successful UNFCCC-related outcomes, the most important question is what role the MEF can or intends to play in terms of securing a comprehensive global climate change agreement by the agreed COP-21 2015 deadline.

Perhaps the important and relevant global outcome of the MEF in terms of the linkage between energy and climate change has been the creation of the Clean Energy Ministerial (CEM), which was first announced by the US Secretary of Energy at the Copenhagen COP in 2009 and was, according to the CEM website (www.cleanenergyministerial.org/about), aimed at "[bringing] together ministers with responsibility for clean energy technologies from the world's major economies and ministers from a select number of smaller countries that are leading in various areas of clean energy". The CEM website also states that there are currently "23 participating CEM governments", which "account for 80 percent of global greenhouse gas emissions and 90 percent of global clean energy investment". The stated focus of the CEM is three global climate and energy policy goals:

- improve energy efficiency worldwide;
- enhance clean energy supply;
- expand clean energy access.

It is listed as having 13 different initiatives that are supposed to support the three goals listed above and which are grouped into three categories (www.cleanenergyministerial.org):

- Initiatives that are working to improve energy efficiency – the Global Superior Energy Performance Partnership and the Super-Efficient Equipment and Appliance Deployment Initiative;
- Initiatives that are working to expand clean energy supplies – Bioenergy Working Group, the Carbon Capture, Use and Storage Action Group, the Multilateral Solar and Wind Working Group and the Sustainable Development of Hydropower Initiative;
- Cross-cutting initiatives that are working to address key issues in clean energy policy, leadership, and access – which are further categorized into "integration" initiatives that are working in areas that cut across traditional energy segments – 21st Century Power Partnership, the Electric Vehicles Initiative, the Global Sustainable Cities Network, and the International Smart Grid Action Network – and "human capacity" initiatives working to build capacity to accelerate the transition to clean energy – the Clean Energy Education & Empowerment Women's Initiative, the Clean Energy Solutions Center (online forum), and the Global Lighting and Energy Access Partnership (LEAP).

The one CEM initiative that has an explicit energy access focus is LEAP, which is stated as working "to transform the global market for affordable, clean, and high-quality off-grid devices for the approximately 1.6 billion people who lack access to grid-supplied electricity" and has an explicit goal of facilitating "access to improved lighting services for 10 million people within five years". According to the CEM, Global LEAP was launched in 2012 and continues the work of the Clean Energy Ministerial's original energy access initiative (the Solar and LED Energy Access initiative), while working to achieve greater impact within the UN SE4All campaign. It is useful to note, however, that within the CEM, there does not appear to be a clear mention of the role of SLCPs such as black carbon (BC) resulting from inefficient cookstoves; nor is the issue of energy access broadened to include all forms of clean energy access services for the poor.

There are a number of initiatives within the CEM on energy. Both the MEF and the CEM are logical fora for addressing the nexus between energy access for the poor and climate change, but it remains to be seen what role the CEM will play when it comes to finalizing global agreement on the post-2015 energy access target within the UN context, and/or global agreement on climate change and SLCP mitigation targets.

The MEF and CEM fora provide an innovative, non-UN-derived context for all major emitters to convene and discuss matters of mutual significance. However, the current global reality is that within the UN context, the UNFCCC's intergovernmental climate change negotiations are tasked with securing a consensus-based global climate change mitigation agreement by 2015. With the time ticking towards a shared 2015 deadline for two separate agreements – one on climate change and the other on a post-2015 development agenda – bold and focused global action will be required to transform existing negotiations silos and their respective outcomes into an integrated framework for action.

There is a broad global consensus that the lack of access to sustainable energy services poses serious development challenges for poorer countries and communities, and that the poor in developing countries, particularly the least developed countries (LDCs) and small island developing states (SIDS0, are the most vulnerable to the adverse impacts of climate change whilst contributing the least in terms of per-capita GHG emissions. At this historical juncture, continuing the practice of having two separate global negotiations silos makes no sense on a variety of levels – intersecting nature of the dual challenges, efficacy of implementation, and waste of limited resources (technical and financial) and time.

Addressing this nexus between lack of energy access for the poor in developing countries and climate change mitigation is fraught with serious political and economic difficulties, especially when concrete action to mitigate GHGs has not been embraced by influential Annex I Parties held responsible under the UNFCCC. What also needs to be clearly acknowledged within the current climate change negotiations with its diminished Kyoto Protocol phase 2 is that the original intent of the UNFCCC's objective of reducing GHG emissions has become politically harder to address via a comprehensive global framework agreement. The new complexities, associated not just with the growth of large-scale aggregate GHG-emitting developing countries, but also with worsening indoor and outdoor air pollution caused by the incomplete and inefficient combustion of traditional fuels and lack of access to sustainable energy services, cannot be simply discounted or avoided by the global community.

Major developing country aggregate GHG emitters, such as China, India, and, to a lesser extent, Brazil and South Africa, are focused on addressing poverty reduction and socio-economic development needs and do not have historical responsibilities currently under the UNFCCC; but their longer-term involvement in climate change is seen as critical to achieving a comprehensive climate change agreement. But the issue that has not received adequate intergovernmental emphasis relates to the costs of worsening air pollution on human health, food security, and the national and regional environment caused by burgeoning aggregate GHGs and other SLCP emissions, particularly in large developing countries. The darkening cloud of indoor and outdoor air pollution facing the poor in developing countries as a consequence of their dependence on inefficient energy services is a new and additional problem that needs to be factored into an already complicated mix of development challenges.

Growing levels of indoor air pollution caused by a lack of access to clean and cost-effective energy services pose the gravest dangers for poor households in developing countries. Over a decade ago, Ezzati and Kamen (2002) focused on the health-related impacts of indoor air pollution from the use of solid fuels in developing countries and pointed to the need for additional research to address implementation needs given that household exposure to indoor smoke due to solid fuel combustion is a complex phenomenon. Equally prescient was an article by Smith (2002), which focused on areas of future research related to indoor air pollution in developing countries, which pointed to the "important potential synergisms between efforts to reduce greenhouse gas emissions and those to reduce health-damaging emissions from solid-fuel stoves" and sounded a cautionary note on the substitution of biomass by coal given "the known serious health effects of household coal use" (p. 198). More recently, Sinha and Nag (2011) have pointed to ill-health caused by air pollution in households in developing countries that depend on biomass fuel (wood, dung, and agricultural residues) and coal as their basic source of energy needs of daily life, for cooking, lighting, and heating. The complex linkages between climate change and worsening indoor air pollution due to heavy reliance on inefficient solid fuels and energy devices and technologies are not easy to address, but the human costs of using unsustainable and polluting energy services is harder to justify in terms of a post-2015 shared development agenda.

In the final analysis, perhaps the most compelling reason for focusing on nationally integrated frameworks for increasing access to sustainable energy services for the poor and reducing SLCPs is derived directly from recent WHO data related to the growing and serious costs of indoor air pollution.

As referenced previously, close to three billion people rely on solid fuels (i.e. wood, charcoal, coal, dung, crop wastes) to cook or heat their homes and use open fires or inefficient, traditional cookstoves, which produce high levels of household (indoor) air pollution (HAP), including pollutants such as soot and fine particulates comprising BC as well as carbon monoxide. According to the WHO, based on most recent 2012 data, 4.3 million people a year die prematurely from illnesses attributable to the HAP caused by the inefficient use of solid fuels. In poorly ventilated dwellings, exposure to HAP due to reliance on solid fuels and traditional practices can exceed acceptable levels for fine particles 100-fold, and exposure is particularly high among women and young children, who spend the most time near the domestic hearth (WHO, 2014b). Table 5.1, based on WHO data, lists the morbidity data related to different diseases associated with indoor or HAP.

The table provides a grim reminder of the health costs borne by the nearly three billion people who each day cook on open fires or crude cookstoves fueled by charcoal, dung, or solid biomass, such as wood and twigs, and suffer the effects of chronic and prolonged exposure to smoke from inefficient

Table 5.1 Global estimates of premature deaths attributable to household air pollution (HAP) caused by inefficient use of solid fuels

Percentage of deaths per disease	HAP exposure related to disease
Pneumonia – 12%	Exposure to HAP almost doubles the risk of childhood pneumonia. Over half of deaths among children less than 5 years old from acute lower respiratory infection are due to particulate matter inhaled from indoor air pollution from household solid fuels.
Stroke – 34%	Nearly one-quarter of all premature deaths due to stroke (i.e. about 1.4 million deaths, of which half are in women) can be attributed to the chronic exposure to HAP caused by cooking with solid fuels.
Ischaemic heart disease – 26%	Approximately 15% of all deaths due to ischaemic heart disease, accounting for over a million premature deaths annually, can be attributed to exposure to HAP.
Chronic obstructive pulmonary disease (COPD) – 22%	Over one-third of premature deaths from COPD in adults in low- and middle-income countries are due to exposure to HAP. Women exposed to high levels of indoor smoke are 2.3 times more likely to suffer from COPD than women who use cleaner fuels. Among men (who already have a heightened risk of COPD due to their higher rates of smoking), exposure to indoor smoke nearly doubles (i.e. 1.9 times) that risk.
Lung cancer – 6%	Approximately 17% of annual premature lung cancer deaths in adults are attributable to exposure to carcinogens from HAP caused by cooking with solid fuels such as wood, charcoal, or coal. The risk for women is higher, due to their role in food preparation.

Source: WHO (2014c).

cookstoves and solid fuels. When combined with the disproportionate economic costs borne by women and girls whose primary responsibility it is to collect traditional solid fuels and cook with inefficient sources and fuel, the proximate costs to human health and well-being in developing countries becomes hard to avoid. But, as discussed in Chapter 1, there is another aspect to the traditional reliance on inefficient cookstoves and solid fuels that needs to be discussed at the global level in an integrated manner – namely, the release of particulate matter less than 2.5 micrometers in diameter (PM 2.5) which is a component of BC emitted as a result of incomplete combustion of solid fuels, biomass including by diesel based sources, cookstoves and open fires.

To be clear and, as discussed previously, AR5 of the Intergovernmental Panel on Climate Change (IPCC), as well as the other global relevant reports, have presented evidence that adverse impacts of climate change will be especially devastating for the poor and vulnerable who ironically contribute the least in terms of actual GHGs per capita to the problem of historical and long-term climate change. If the Durban Platform's objective to scale up mitigation efforts is to be realized by COP-21 in 2015, then bold action related to the global climate change negotiations cannot dismiss recent scientific analyses related to emissions reductions actions that focus on near-term strategies such as those that can be achieved by addressing SLCPs (also commonly referred to as short lived climate forcers; SLCFs) – amongst them BC particles, tropospheric ozone, methane and some HFCs (UNEP/WMO, 2011).

The confluence of development-related challenges associated with lack of access to sustainable, and cost-effective access to energy services for the poor needs to be more clearly recognized within the UN context and scaffolded by private–public partnerships and voluntary initiatives. But this nexus is a complicated and challenging one to address, because it goes to the core of poverty reduction and economic development efforts and hence can be a source of political tension in terms of implementation. This nexus is also extremely difficult to address in terms of climate change mitigation, because the objective of the UNFCCC is quite clear in ascribing principal and historical responsibility for GHG emissions to developed/Annex I Parties, and hence developing countries, including large-scale aggregate GHG emitters, have proved to be pointedly reluctant to accept any expanded responsibility for mitigation action. But what an integrated framework on energy access for the poor and climate change is forced to confront is that continuing to ignore the serious socio-economic and environmental costs borne by the poor due to their reliance on inefficient and polluting energy services is no longer viable or justifiable in terms of a shared post-2015 development agenda or a comprehensive global climate change mitigation agreement.

It is worth reiterating that the emissions reductions of these SLCPs/SLCFs cannot substitute for the agreed-upon emissions reductions of GHGs such as carbon dioxide, which are required to limit longer-term climate change impacts, nor are they to be viewed as a mechanism to shift the historical responsibility for anthropogenic climate change to developing countries. However, what is equally important to recognize is that discounting the regional and national benefits to health, agriculture, and food security that accrue from any future nationally based and regionally relevant actions related to reducing emissions of key SLCPs would be a missed opportunity for all, especially for those developing countries where any such context-specific and nationally driven actions would have the most benefits. In other words, the idea of having regionally derived and nationally driven integrated frameworks for action that reduce indoor airborne pollutants that are also SLCPs resulting from the inefficient solid fuels and cookstoves, including through increased access to innovative energy technologies, is extremely worthwhile, because such frameworks offer joint sustainable development and climate benefits. Such frameworks would have be customized based on national and regional needs and concerns, but could provide multiple positive impacts on human health, livelihoods, and socio-economic and agricultural productivity; allow for a broader range of participation and involvement by interested countries in resolving both climate change and energy for sustainable development; and leverage the much-needed engagement of a diverse range of stakeholders.

Burning solid fuels such as traditional forms of biomass is inefficient, and when these solid fuels are burnt using ineffective cooking and heating devices, they release a toxic mix of health-damaging and environmental pollutants such as BC that have short life spans but have significant impacts to climate change at regional and global levels. BC exists as particles in the atmosphere and is a major component of soot emitted from many common sources, including cars, trucks, residential cookstoves, and forest fires. To be clear, BC:

> ...results from the incomplete combustion of fossil fuels, wood and other biomass.... The black in BC refers to the fact that these particles absorb visible light. This absorption leads to a disturbance of planetary radiation balance and eventually leads to warming. The contribution of warming of 1 gramme of BC seen over a period of 100 years has been estimated to be anything from 100 to 200 times higher than a gramme of CO_2. An important aspect of BC particles is that their lifetime in the atmosphere is short, days to weeks, and so emission reductions have an immediate benefit for climate and health. (UNEP/WMO, 2011, p. 6)

Several additional studies have focused on BC as a short-lived atmospheric fine particulate matter that results from the incomplete combustion of fossil fuels and biomass, and have discussed its role as the strongest absorber of solar radiation in the atmosphere, including its contribution to climate change (Forster *et al.*, 2007; Ramanathan and Carmichael, 2008). The role of BC in contributing to disruptions in rainfall patterns, including the monsoons in South Asia, has been examined by Ramanathan *et al.* (2005), as well as Lau *et al.* (2008), while its linkages to heating of glaciated regions in the Himalayas has been discussed by Flanner *et al.* (2009) and Menon *et al.* (2010). Meanwhile, the impacts of SLCPs emanating from road transport-related pollution on the Himalayas has been documented by Marioni *et al.* (2013)). According to Ramanathan *et al.* (2005): "South Asian emissions of fossil fuel SO_2 and black carbon increased 6-fold since 1930, resulting in large atmospheric concentrations of black carbon and other aerosols" and "this period also witnessed strong negative trends of surface solar radiation, surface evaporation, and summer monsoon rainfall" which raises "the possibility that, if current trends in emissions continue, the subcontinent may experience a doubling of the drought frequency in the coming decades" (p. 5326).

Given the population of the poor in South Asia, and their heavy dependence on monsoons for food security, the possibility of increased frequency of droughts poses severe development challenges. Meanwhile, the strong linkages between BC emissions and biomass and fossil fuels usage patterns in rural India (Indo-Gangetic Plain) and the impacts of the same were the subject of a detailed field study by Rehman *et al.* (2011). This field study found that previous health impact and climate impact studies of BC had underestimated the risks and pointed out that "the atmospheric solar heating as well as the surface dimming over the Indo-Gangetic Plain due to BC and other particles from cook stove emissions should be larger by factors ranging 2 or more" and that "understanding of BC effects on monsoon and Himalayan glaciers needs to undergo a major revision". Most significantly, the study concluded by stating:

> Considering the long-term actions required for the mitigation of CO_2 and related pollutants, the most effective strategy that would yield dividends both for environment and health in a short time span is to focus on short lived carbonaceous pollutants such as BC. The strategy would be particularly relevant for developing countries where the reduction of short-lived pollutants by introduction of efficient technologies for burning biomass for cooking is a development imperative. (Rehman *et al.*, 2011, p. 7297)

In terms of any future integrated framework for action on improving sustainable energy access for the poor and mitigating against SLCPs/SLCFs, perhaps the most influential global report to date has been the joint 2011 UNEP/WMO assessment. This joint assessment report has provided a crucial impetus to the global debate on the role of addressing SLCPs/SLCFs and resulted in the formation of the CCAC, which was discussed earlier. More recently, the issue of SLCPs was raised at the Ascent meeting – the sole precursor meeting to the 2014 Climate Summit. The key findings of the report underpin not only the work of the CCAC but also any anticipated "bold announcements" that relate to the UN Secretary-General's identification of SLCPs as one of the nine key action areas in the upcoming Summit with direct ramifications for any proposed 2015 climate agreement.

Prepared jointly by UNEP and WMO, the assessment convened 50 experts and resulted in the identification of a package of 16 policy measures that were found to be most effective in achieving large emissions reductions of SLCPs. The Assessment found that measures to reduce SLCFs, implemented in combination with carbon dioxide control measures, would increase the chances of staying below the 2°C target agreed to by the Parties to the UNFCCC in 2010 and listed the following benefits:

- Full implementation of the identified measures would reduce future global warming by 0.5°C (within a range of 0.2–0.7°C). The rate of regional temperature increase would also be reduced.
- Both near term and long-term strategies are essential to protect climate. Reductions in near term warming can be achieved by control of the short lived climate forcers, whereas carbon dioxide emission reductions, beginning now, are required to limit long-term climate change.
- Full implementation of the identified measure would have substantial benefits in the Arctic, the Himalayas and other glaciated and snow-covered regions. This could reduce warming in the Arctic in the next 30 years by about two-thirds compared to the projections of the Assessment's reference scenario. This substantially decreases the risk of changes in weather patterns and amplification of global warming resulting from changes in the Arctic.
- Full implementation of the identified measures could avoid 2.4 million premature deaths and the loss of 52 million tonnes, 1–4 percent of the global production of maize, rice, soybean and wheat each year. The most substantial benefits will be felt immediately in or close to the regions where action is taken to reduce emissions, with the greatest health and crop benefits expected in Asia (UNEP/WMO, 2011, p. 3).

One of the most important points to consider is that the Assessment's listing of benefits found that implementation of response measures would accrue and be felt most significantly on a regional and national basis, which lends further support for nationally driven and regionally relevant frameworks for integrated action to improve access to sustainable energy services for the poor and also reduce BC emissions. According to the assessment, at the national level "many of the identified measures could be implemented under current and existing policies designed to address air quality and development", and improved cooperation at the regional level "would enhance the widespread implementation and address transboundary climate and air quality issues". These findings of the assessment, along with its stated confidence that "immediate and multiple benefits will be realized upon implementation of the identified measures", are worth underscoring in terms of addressing linkages between climate change and energy-related air quality concerns (UNEP/WMO, 2011, p. 4).

Table 5.2 is excerpted from the UNEP/WMO assessment and lists the 16 different policy measures. It is worth highlighting because it represents an

Table 5.2   Measures (16 methane- and black carbon [BC]-related) that improve climate change mitigation and air quality and have large emission reduction potential (measures related to BC and residential sector usage in developing countries are in bold)

Measure	Sector
CH$_4$ (methane) measures	
Extended pre-mine degasification and recovery and oxidation of CH$_4$ form ventilation air from coal mines	Extraction and transport of fossil fuel
Extended recovery and utilization, rather than venting of associated gas and improved control of unintended fugitive emissions from the production of oil and natural gas	
Reduced gas leakage from long-distance transmission pipelines	
Separation and treatment of biodegradable municipal waste through recycling, composting, and anaerobic digestion, as well as landfill gas collection with combustion/utilization	Waste management
Upgrading primary wastewater treatment to secondary/tertiary treatment with gas recovery and overflow control	
Control of CH$_4$ emissions from livestock, mainly through farm-scale anaerobic digestion of manure from cattle and pigs	Agriculture
Intermittent aeration of continuously flooded rice paddies	
BC MEASURES (affecting BC and other co-emitted compounds)	
Diesel particle filters for road and off-road vehicles	Transport
Elimination of high-emitting vehicles in road and off-road transport	
Replacing coal by coal briquettes in cooking and heating stoves	
Pellet stoves and boilers, using fuel made from recycled wood waste or sawdust, to replace current wood-burning technologies in the residential sector in industrialized countries	Residential
Introduction of clean-burning biomass stoves for cooking and heating in developing countries	
Substitution of clean-burning cookstoves using modern fuels for traditional biomass cookstoves in developing countries	
Replacing traditional brick kilns with vertical shaft kilns and Hoffman kilns	Industry
Replacing traditional coke oven with modern recovery ovens, including improvement of end-of-pipe abatement measure in developing countries	
Ban of open field burning of agricultural waste	Agriculture

Source: UNEP/WMO (2011, p. 9).

immediate set of menu options related to the achievement of large emissions reductions of SLCPs/SLCFs. The 16 measures identified – seven aimed at reducing methane emissions and nine aimed at reducing BC emissions – were chosen from a larger set of 2,000 measures, because they are seen as delivering more than 90% of the global benefit of the total of 2,000 measures modeled (UNEP/WMO, 2011, p. 8).

These measures and additional approaches to HFC reduction have also informed a review of the World Bank's extensive energy sector portfolio, which has identified and highlighted potential opportunities and activities that could reduce emissions of SLCPs (Sameer *et al.*, 2013). Table 5.2 provides a summary list of 16 methane- and BC-related measures that could improve air quality and have large emission reduction potential. It is crucial to note, however, that only a subset of these 16 measures, that is only four specific measures emphasized in Table 5.2, are related to residential and household energy services and only three of these are of direct relevance to developing countries, with an even smaller number – two (bold emphasis added) – related to use of cookstoves that are predominant in poor and low-income households. Nevertheless, this subset of measures is worth highlighting for its significance in terms of jointly addressing concrete energy access needs and BC reductions in developing countries.

There is no escaping the fact that at the global level, the energy sector plays an absolutely fundamental role in any current or future global response related to anthropogenic global climate change, as the energy sector is the single largest source of anthropogenic GHG emissions. Consequently, it can be, and has been, argued that limiting these emissions should be the principal locus of global action in terms of climate change mitigation. But access to safe and reliable energy (sources, services, and systems/technologies) drives human development in all countries around the world, and conversely lack of access to energy impacts negatively on human well-being, and adversely impacts on the lives of the poor. The global community has a unique opportunity to look afresh at the intergovernmental negotiating processes related to climate change and energy for sustainable development with a view towards highlighting those efforts that lend themselves to synergistic action and partnership. The operating premise is that in order to enhance and scale up action related to climate change for developing countries, addressing policy linkages between the intergovernmental process on energy for sustainable development and that of global climate change is a necessary first step.

An examination of key global outcomes emanating from global climate change and sustainable development processes convened by the UN reveals a missing linkage to energy access for the poor in both sets of negotiated outcomes, and it is this linkage that is crucial from the broader and urgent perspective of arriving at a shared 2015 development agenda and a robust climate change agreement that involves all large-scale GHG emitters. Key globally agreed outputs of 20 years of UN-led global climate change negotiations and 40 years of UN-led environment and sustainable development negotiations reflect a puzzling disconnect, in that increasing access to sustainable energy services for the poor and climate change have not been addressed in an integrated manner. It is increasingly hard to justify having two separate negotiations tracks on sustainable development and climate change, precisely because of the broad UN consensus emerging from the 2012 FWW that a shared development agenda is needed for the post-2015 era. It is indeed time for a single streamlined agenda, because increasing access to sustainable energy services for the poor is critical to addressing both poverty reduction and climate change. Avoiding the nexus between energy access for the poor and climate change quite simply appears to be at odds with broader UN rhetoric and the push towards an integrated and universal 2015 development agenda that puts the goal of poverty reduction at front and center.

Long-term climate mitigation will only be possible if deep and persistent cuts in CO_2 emissions are made. Fast action to reduce SLCPs represents an area for multiple development benefits in the areas of health, energy and short-term climate benefits, but should not be "seen as a substitute for (or an opportunity to delay) urgent mitigation of CO_2 emissions but rather as complementary action" (UNEP, 2013, p. 10). Measures that can address dual intersecting concerns of improving indoor air pollution in poor households and climate change mitigation of SLCPs should be implemented as an urgent development imperative, because they offer dual and intersecting benefits that are directly relevant to the nations and regions where they are implemented. Given the national and regional nature of impacts of SLCPs on climate, health, food security and agriculture, and ecosystems, it is only logical that regional and national approaches that are context-specific and responsive to national and regional could be undertaken.

In 2009, the Convention on Long-Range Transboundary Air Pollution (CLRTAP), which covers Europe, Central Asia, and North America, recognized that BC and organic carbon are important components of particulate matter (PM), and that exposure to fine particles leads to tens of thousands of premature deaths and serious health effects in the UN Economic Commission for Europe (UNECE) region. The Executive Body of the CLRTAP established an Ad Hoc Expert Group on Black Carbon (EGBC) to assess available information on black carbon and to identify options for potential revisions to the Gothenburg Protocol that would enable the Parties to mitigate BC as a component of PM (UNECE, 2010). In 2011, CLRTAP agreed to address BC in its revision of the Gothenburg Protocol and to consider the impacts of methane in the longer term. The potential role of other regional agreements such as the Male Declaration on Control and Prevention of Air Pollution and its likely Transboundary Effects for South Asia, the Association of Southeast Asian Nations HAZE Protocol, the Lusaka Agreement, the Nairobi Agreement, and the Abidjan Agreement are all useful starting points for potential regionally based and nationally driven integrated action that can simultaneously increase energy access for the poor and also address climate concerns. The multiplier benefits of reducing SLCPs can and should be contrasted with the multiplier threats faced by the poor due to their heavy reliance on inefficient energy sources and systems. Reducing emissions of SLCPs, in particular BC, represents a major opportunity to deliver multiple benefits in terms of public health, food, and energy security and near-term climate protection and should not be discounted or ignored.

The global community has a unique opportunity to look afresh at the inter-governmental negotiating processes related to climate change and energy for sustainable development, with a view towards highlighting those efforts that lend themselves to synergistic action. Inefficient household energy use has adverse consequences for the environment, air quality, and human health. Emissions from traditional cookstoves also contribute significantly to outdoor air pollution and exacerbate already deadly air pollution in large towns and cities around the world, affecting those with and without access to clean household energy. Integrated action on increasing access to sustainable energy services for the poor by replacing polluting and ineffective energy services, and also thereby reducing BC and soot emissions, is one example of practical and integrated action on energy access and climate change mitigation of SLCPs.

Major constraints need to be overcome which have not been discussed, including but not restricted to the point that substantial investments by governments, global donors, and the private sector will be needed to scale up access to modern energy services – in particular, targeted financial and capacity-building support for the development of appropriate technologies, services and infrastructure. Bold action will require breaking through long-standing silos on UN-led intergovernmental action on climate change and energy access for the poor. The silo approach to energy and climate change, whereby the global climate change negotiations are distinct and separate from global negotiations on energy for sustainable development, has led to the puzzling absence of concrete linkages between energy and climate change objectives that are the principal driver for national mitigation efforts. This absence of a concrete nexus between climate change and energy for sustainable development at the level of intergovernmental negotiations impacts on the post-2012 climate change framework because it limits practical scalable emissions reductions that derive directly from nationally driven energy sector needs and concerns.

The point that energy for sustainable development objectives are largely missing from globally agreed outcomes on climate change; and that energy access for the poor remains largely unaddressed in the terms of specific targets and goals in key globally agreed outcomes on sustainable development and climate change are surprising yet relevant and actionable findings for any post-2105 development agenda. The overall lack of programmatic linkages between energy access for the poor and climate change mitigation may not surprise veterans of the climate change negotiations process, who have extensive experience in parsing concrete references and/or possible targets related to energy sector mitigation, but it does a disservice to the broader UN quest for a shared post-2015 development agenda anchored by an overarching goal to reduce poverty and address climate change. In this regard, the reduction of SLCPs such as BC is especially important to highlight when considering the linkages between anthropogenic global climate change as well as the human health impacts that the poor have to face due to their reliance on polluting energy services.

Bridging the divide between energy, poverty reduction, and climate change objectives is critical to large developing countries that also happen to be increasingly large aggregate GHG emitters dealing with pressing energy access needs. Unprecedented levels of global commitment will be needed to leverage an integrated and responsive framework for action on sustainable energy for the poor and climate change mitigation. Regionally and/or nationally driven integrated frameworks for action that increase access to sustainable energy services and simultaneously allow for reduction of SLCPs are areas worth exploring because such frameworks could directly address the nexus between climate change and energy access for the poor in developing countries.

Putting energy sector issues at the front and center of the global climate change negotiations enables the active engagement of private sector and civil society actors, and the formulation of innovative partnerships for joint action on energy access and climate change, including action on SLCPs, which have been largely absent from the global climate change negotiating framework, but which are significant for any future global climate change agreement. In this regard, proposals for a possible tropospheric ozone protocol, or regionally derived and nationally driven framework agreements on BC, should be considered in terms of promoting innovative partnerships and immediate scalable action at the country level by interested countries and actors. Energy for sustainable development objectives that are crafted nationally yet relevant globally and regionally, including energy access targets that are clearly linked with national climate change mitigation objectives, offer promising opportunities for bold action. In the near future, "bold action" lies in comprehensive linkages between energy for sustainable development objectives and climate change mitigation goals. Expanding these linkages with concrete programmatic and policy guidance in 2015 can provide a means for engaging all relevant stakeholders, ranging from national governments to civil society actors, to scale up plans, programs, and policies and transform long-standing global silos on energy access and climate change into the actual delivery of sustainable energy services for the poor.

In the final analysis, ignoring the regional and national benefits to human health, environment, and energy security that accrue from concrete action linking increasing energy access for the poor targets with climate change objectives based on proven strategies would be a missed opportunity not only for the broader UN quest for a shared 2015 development agenda, but especially for those developing countries where these nationally driven actions would have the most immediate and lasting benefits. The idea of exploring new initiatives that integrate energy access and reduce SLCPs, such as BC, is not only urgent but also effective from a broader global development perspective. The role of innovative energy services and technologies for the poor, and regionally derived and nationally driven frameworks for action that jointly address the growing health and morbidity burden on the poor due to inefficient and insufficient access to energy and also limit BC emissions are areas worth exploring. Within the UN-led global context, the time and resources needed to address the massive development challenges of reducing poverty by increasing access to sustainable energy services and also mitigating climate change are gravely constrained. It makes sense, therefore, to transform erstwhile silos on energy for sustainable development and climate change that consign the poor to living in misery and ill-health into an integrated framework for post-2015 development action.

References

CCAC website. About CCAC. http://www.unep.org/ccac/ [accessed May 19, 2014].

CCAC (2014) *Executive Summary: 2014.* http://www.unep.org/ccac/Portals/50162/docs/ccac/WG-APR2014-2-%20CCAC_Executive_Summary.pdf [accessed May 19, 2014].

Ezzati, M. and Kamen, D.M. (2002) The health impact of exposure to indoor air pollution from solid fuels in developing countries: Knowledge, gaps and data needs. *Environmental Health Perspectives*, 110(11), 1057–1068.

Flanner, M.G., Zender, C.S., Hess, P.G. *et al.* (2009) Springtime warming and reduced snow cover from carbonaceous particles. *Atmospheric Chemistry and Physics*, 9, 2491–2497.

Forster, P., Ramaswamy, V., Artaxo, P. *et al.* (2007) Changes in atmospheric constituents and in radiative forcing. In: Solomon, S., Qin, D., Manning, M. *et al.* (eds) *Climate Change 2007: The Physical Science Basis. Contribution of Working Group 1 to the Fourth Assessment Report of the IPCC.* Cambridge: Cambridge University Press.

GACC website. Global Alliance on Clean Cookstoves: Alliance Mission and Goals. http://www.cleancookstoves.org/the-alliance/ [accessed May 15, 2014].

Lau, K.M. Ramanathan, V., Wu, G.-X. *et al.* (2008) The joint aerosol-monsoon experiment: a new challenge for monsoon climate research. *Bulletin of the American Meteorological Society*, 89, 369–383.

MEF (2009) *Declaration of the Leaders of the Major Economies Forum on Energy and Climate.* http://www.majoreconomiesforum.org/past-meetings/the-first-leaders-meeting.html [accessed April 27 and September 24, 2014].

Marinoni, A., Cristofanelli, P., Laj, P. *et al.* (2013) High black carbon and ozone concentrations during pollution transport in the Himalayas: Five years of continuous observations at NCO-P global GAW station. *Journal of Environmental Sciences*, 25(8), 1618–1625.

Menon, S., Koch, D., Beig, G. *et al.* (2010) Black carbon aerosols and the third polar cap. *Atmospheric Chemistry and Physics*, 10, 4559–4571.

Ministry of Information and Broadcasting (2014) Text of Statement made by MoS for Environment, Forests and Climate Change Shri Prakash Javadekar, at United Nations Climate Summit 2014. http://inbministry.blogspot.com/2014/09/text-of-statement-made-by-mos-for.html [accessed October 13, 2014].

PCIA legacy website. Partnership for Clean Indoor Air. http://www.pciaonline.org/ [accessed May 3, 2014].

Ramanathan, V. and Carmichael, G. (2008) Global and regional climate changes due to black carbon. *National Geoscience*, 1, 221–227.

Ramanathan, V., Chung, C., Kim, D. *et al.* (2005) Brown clouds: impacts on South Asian climate and hydrological cycle. *Proceedings of National Academy of Sciences*, 102(15), 5326–5333.

Sameer, A., Ebinger, J., Kleiman, G. and Oquah, S. (2013) *Integration of Short-lived Climate Pollutants in World Bank Activities: a Report Prepared at the Request of the G8*. Washington, DC: World Bank. http://documents.worldbank.org/curated/en/2013/06/18119798/integration-short-lived-climate-pollutants-world-bank-activities-report-prepared-request-g8 [accessed May 19, 2014].

Sinha, S.N. and Nag, P.K. (2011) Air pollution from solid fuels. In: Nriagu, J., ed., *Encyclopedia of Environmental Health*. Oxford: Elsevier.

Smith, K. R. (2002), Indoor air pollution in developing countries: recommendations for research. *Indoor Air*, 12(3), 198–207.

UN (2013) *The New Global Partnership: Eradicate Poverty and Transform Economies through Sustainable Development*. New York: UN.

UN (2014a) "UAE to Host High-Level Meeting in May Leading up to the UN Secretary-General's Climate Summit", February 3, 2014. http://www.un.org/climatechange/summit/2014/02/04/uae-to-host-high-level-meeting-in-may-leading-up-to-the-un-secretary-generals-climate-summit/ [accessed May 7, 2014].

UN (2014b) "Climate summit to open opportunities for world leaders to show new actions", February 10, 2014. http://www.un.org/climatechange/summit/2014/02/10/climate-summit-to-open-opportunities-for-world-leaders-to-show-new-actions/ [accessed May 7, 2014].

UN Climate Summit 2014/Executive Office of the UN Secretary-General (2014) Build Consensus and Implement Actions For a Cooperative and Win-Win Global Climate Governance System Sept 23, 2014, Statement by H.E Zhang Gaoli. http://statements.unmeetings.org/media2/4628014/china_english.pdf [accessed October 1, 2014].

UNECE (2010) *UNECE Convention on Long-range Transboundary Air Pollution Aims to Reduce Black Carbon Emissions*. http://www.unece.org/press/pr2010/10env_p20e.html [accessed April 24, 2014].

UNEP (2013) *UNEP Year Book: Emerging Issues in Our Global Environment*. Nairobi: UNEP.

UNEP/WMO (2011) *Integrated Assessment of Black Carbon and Tropospheric Ozone: Summary for Decision Makers*. Nairobi: Kenya. http://www.unep.org/dewa/Portals/67/pdf/BlackCarbon_SDM.pdf [accessed April 24, 2014].

UNGA (2012) *The Future We Want*. Resolution adopted by the General Assembly on 27 July 2012. A/RES/66/288. New York: UN.

US State Department (2012) *Fact Sheet: The Climate and Clean Air Coalition to Reduce Short-Lived Climate Pollutants*. http://www.state.gov/r/pa/prs/ps/2012/02/184055.htm [accessed May 19, 2014].

White House/Office of the Press Secretary (2014) Remarks by the President at the U.N. Climate Summit. http://www.whitehouse.gov/the-press-office/2014/09/23/remarks-president-un-climate-change-summit [accessed October 13, 2014].

WHO (2014a) "7 million premature deaths annually linked to air pollution". News Release. http://www.who.int/mediacentre/news/releases/2014/air-pollution/en/ [accessed April 22, 2014].

WHO (2014b) *Household Indoor Air Pollution*. http://www.who.int/indoorair/en/ [accessed April 23, 2014].

WHO (2014c) *Household Air Pollution and Health*. Fact sheet No. 292. http://www.who.int/mediacentre/factsheets/fs292/en/ [accessed May 11, 2015].

Yeo, S. (2014) Ban Ki-moon: China must offer global climate leadership. RTCC. http://www.rtcc.org/2014/05/20/ban-ki-moon-china-must-offer-global-climate-leadership/ [accessed August 15, 2014].

Index

Note: page numbers in *italics* refer to figures, those in **bold** refer to tables and boxes.

Energy and Global Climate Change: Bridging the Sustainable Development Divide,
First Edition. Anilla Cherian.
© 2015 John Wiley & Sons, Ltd. Published 2015 by John Wiley & Sons, Ltd.